FOREWORD

10.22334. A grain of sand.

Variothoughts is sand, not gold. According to Syntalism, our world is built of sand, not gold. The most valuable thing is sand, sand is bread, and gold is salt.

10.22339.

There are no questions that cannot be answered in the Variothoughts. The Variothoughts is an endless source of inspiration for hearts searching for truth.

10.22266. Nutcracker.

Breaking stereotypes and patterns. Variothoughts is a brain-crushing book, the meaning of which is to achieve the integrity of the mind. First, all beliefs should be destroyed in the dust, and then it will all stick together and enlightenment will come.

10.22554. The living and the dead book.

In the original, Variothoughts is the ideal of the perfection of truth, but the ideal is dead and, therefore, there is no joy in it. To bring back the joy of life to Variothoughts, I decided to salt the bread. Salt-free bread is too sweet. In Variothoughts translations, I threw a couple of spoons of chaos. My act of monstrous vandalism led to the loss of 20% of the meaning, and made the texts very strange and obscure. You will call me a scoundrel and a vandal, but I don't think so ... On the contrary, I believe it made the texts charming, created artificial barriers ... Now, to understand the texts of Variothoughts and find the truth, you have to smash your head and think. Thinking is joyful.

10.15530.

The basis of speech is truth. Cognition of truth should begin with clarifying the meaning of words.

10.16833.

The main feature of Variothoughts is its unprecedented honesty. No censorship of thoughts, absolute freedom of ideas and words.

5.412.

Variothoughts is a book for those who save their time. Ready-made Lego cubes used to put together any ideas and goals. The DNA and RNA of thought.

3.2212. Attainment of truth.

A comprehensive attainment of reality occurs by knowledge's thinning of its tiniest degree of detail.

3.2213.

Unity is hidden in differences. You unite by disuniting. Holding one onto another, the infinitely small becomes the infinitely big.

10.6168.

The Variothoughts is a basic library of DNA of thought, loading it into the brain can solve any problem. Any dreams, any goals will be available to you, thanks to the philosophy of Syntalism.

3.1971.

Variothoughts implements divergent thought algorithms in order to come again to unity through a multitude. Many grows out of one and one grows again out of many.

3.2211.

Soundness is the ability to dynamically examine things from

different perspectives.

3.2216.

Having reached its limit, knowledge transforms into will. What is the limit of knowledge? – Faith.

10.21243. Love of truth.

The meaning of human life is to overcome infinite loneliness and find infinite love.

THE UNIVERSE. RATIONALITY PRINCIPLE

356.

Taking into account that matter is information about the form and structure of energy and the fact that the universe expands ...there can be drawn a conclusion that the Universe is an artificial intelligence system that endlessly accumulates information and hence expands...

3.92.

Interesting. Expansion of the universe after the big bang and aiming of the order to chaos are evidently the same processes. Creation of the universe is basically destruction of it.

3.93. After the explosion.

Degradation can go on and on and the very fact that our universe exists proves it.

3.443.

The dualism of the wave and corpuscular nature of light is due to the fact that in a positive wave a quantum is created and a particle exists, and in a negative wave a particle is destroyed and energy is released. That is, a particle is created and destroyed, created and destroyed.

412.

Hear the hum of the Universe -
countless bees ... the children of it -
turn a wheel of life...

3.461. Perpetual motion.

The wave structure of being is something like a taut bowstring or cocking a spring.

476.

The Universe consists of zeroes and sometimes of ones.

1096.

The Universe is a big self-teaching system, the goal of existence of which is about betterment and creation of new forms of information. Thus, there are mechanisms which fulfil this goal.

677. Axiom of Eggs and Chicken [In brevi]

Here comes an interesting conclusion about one-moment existence of everything and every thing.

Everything - the Universe, time and space exist in a single close moment, inside which there already exists its relative time, like a point in a system of coordinates.

In essence, the understanding of this mechanism helps to understand the concept of an endless point inside which there exists a huge Universe...

The question is only in the effectiveness of codes in the energy of information about its structure...

3.442.

The essence of the wave is the spring and the accumulation of kinetic energy of time. In the negative half-wave time, like a spring, is inhibited and energy accumulates due to destruction, in the positive half-wave stored energy creates an explosive effect.

10.6969.

To create the universe, it must be destroyed. To create a new self, the old self must be destroyed.

709. [In brevi]

Anything that once was created will live forever. It's only the shell that can be ruined, but the inner essence will exist. What's more the same goes with everything else: people, things and even thoughts. The Universe is information and this information is saving.

10.5515.

Energy, form, time. Form enslaves energy, and time frees energy. Form is space limited by time.

752. [In brevi]

An absolute point. The point where one moment exists.

When in a single moment there exists all time and when in a single point there exists all space. That's approximately what our Universe looks like from side.

10.5514.

The real world is energy limited by form. Form is an entity limited by time.

941.

On a spacetime scale many things seem really tiny.

10.5224. Nothing and Nothing.

There is only energy and information. Information is an illusion. But I'm telling you, there's nothing but Nothing and information. Energy is also an illusion.

10.4964.

Chaos consists of order. Chaos is a lot of order. The greater the order, the greater the chaos ... Chaos is a superorder.

2278. Every person should control feelings.

People tend to be self-destructive. The Universe tends to fall to pieces.

3.173. A one million light years step.

Quantum leap technology allows to fold universe into a point, a ball... a big ball... a medium sized ball and then unfold it back.

3246. The balance.

The Universe of existence and nonexistence.

3.413.

Although you are terribly afraid of uncertainty, it must be admitted that uncertainty does not exist. Chance is an illusion and a lie.

3.414.

The universe is vain, it grows and does not want to stop.

3.415.

Black matter is the realization of the wave principle of creation.

3.421.

Chaos is a very fast movement, it allows the chaos to save and accumulate a huge amount of energy that can be removed, it slowed down, turning to the right.

3.428.

Order does not do anything special, except that it causes chaos to move not chaotically, but according to certain algorithms. Created algorithms. The order is a information, giving chaos form.

3.434.

What is creation? Self-destruction?

3.437. Slow down.

It just seems like the order increases the speed, in fact, it reduces it, but gives it a vector. The speed of chaos is higher than order, but it is random and has no purpose. Order is the purpose and inhibition of chaos.

3.440.

Chaos is the highest form of order. Chaos adult corporado. A more simplified and delayed form of chaos is called order. Order does not collapse into chaos, it grows into it. Chaos, on the contrary, is destroyed in order.

3.444.

The nature of time has the same nature of dualism as light. Time is both a particle of matter and a radiation of energy. An ultra-high-frequency oscillation, where every moment our world is destroyed and created anew.

3.459.

The smaller the substance, the more it has tension and, consequently, energy in it.

10.22219.

Time and space are one in the sense that space is the result of the disintegration of the one-dimensional essence of time into a three-dimensional essence. In the process of this decay, energy is released from which our world is created.

10.22218.

Time and space is a system of form and content, where space is information and time is energy.

10.22206. The electric spirit pierces the darkness.

Combining the point of view of religion and physics, we can say that the God who created our universe, it was a quantum fluctuation of the energy field that gave energy to the debt of our universe. With this money, our universe was built, and then the debt was paid and God went on to create new universes. In fact, God is a wandering energy field.

10.22215.

What is time? The process of expanding the past at the expense of energy coming from the future, that is, from outside. Does the universe grow like this? Time and space are identical entities, so we can assume that the expansion of the universe occurs in a similar way. The universe is a system that absorbs energy from the outside, and so expands by dividing into internal forms.

9.2.

Question. Does the idea that money can be printed violate the energy conservation law?

Question. Does the fact that the Universe appeared from a point and started to expand mean that the energy creation process is nevertheless going on?

Question. Is the acceleration of the Universe expansion the source of its energy?

Question. Is the Universe expansion mechanics linked with the idea of the unity and struggle of opposites and is it a source of energy?

Question. Money is energy, solar energy is energy, too. Can we say that the process of printing money is the process of the solar energy conversion into the money form and it is linked with the number of people and infrastructure in the state through derivatives?

Question. Can we say that the faster a society grows, builds and develops, the more money the state can print? In the notion of the link between the amount of the available energy to the society growth acceleration.

Question. If the growth has stopped, can the excess money supply be simply burnt thus making the energy conversion? Energy can exist in various forms. Money is a form of energy. Debts are also a form of energy, a form of money. If the dead debts, accumulated in the banking system, are burnt, it will undoubtedly be the act of energy conversion and release.

640.

The universe is a super complicated system trying to become aware of itself; in essence it's that very human child,

that has a long way to perfection and uniqueness ahead of him.

675.

An open question also remains about whether information can exist outside this Universe and according to which principles it exists over there...

... about the role of this Universe sees as some information medium...

There's a high probability there are a lot of such mediums. They are created and erased and this initializing code is transmitted between them in a particular way, - resembling the birth of a human soul... It is also born with some basic set of rules and later fulfils its aim and predestination.

678. [In brevi]

Beyond the limits of the Universe there are no coordinate constants of this Universe as well as there is no space and no time. Hence the Universe, in essence, exists in a single endless point without time and space.

Perhaps, the very creation of the Universe was connected to some event. Let's call it some informational virus that structured the information about the form of the energy so that it after having produced 4 existing dimensions got an incentive for an endless growth...

like as if it opened ...

like as if there was that very Big bang...

9.12.

In particular cases, the energy conservation law does not take into account the "Time" variable. There is an increase in the amount of energy in time in our Universe. The older the Universe, the more energy it has.

9.67. The last God alive.

Turning the society from the three-class one into the two-class one is degradation, as three-phase systems are more energy-efficient than two-phase ones. Moreover, class contradictions were an important source of progress, if they are reduced, the society will start degrading. Furthermore, when the elites replace the lower classes by robots and artificial intelligence systems, the issue of what to do with the majority of people will arise.

The answer will be evident – to immerse them into illusive dream to make them perish in sweet pleasures, like a fly in the syrup. This decision will turn all the social elevators and ladders off. The renewal of blood and thought among elites will stop. That is, interclass motion will disappear along with classes. And motion, as we remember, was the basis of life and an energy access mechanism. Anyway, everything will end and perish rather quickly. In a couple of thousands years the last God alive will create a new Universe and, having erased his memory, will self-destroy.

9.74. Nothing.

It is not the Universe expanding, pushed from within by the energy of the big bang, but it's the void that requires filling, drawing it into itself. Gravity of the darkness attracts light at the speed of darkness. Infinite desire is Nothing, becoming something is the main source of movement in this Universe.

1479.

The task of parents is to raise uniqueness and perfection within their children. Parents are the creators of their children, the creators of the new world. And it's purely their responsibility for how unique and perfect this new Universe will be. It's important to raise wisdom and power within children.

The task of parents is to bring up, teach and train children, making them stronger, wiser and more adaptable to life; introduce children to basic laws of the world, ethics and morality. In essence, the task of parents is to look back on their life and mistakes and try to bring up a new better and stronger generation while taking into account their own experience.

2121. The concept of inkblot.

The universe resembles an inkblot...

Cosmic chaos... absolute novelty and eventuality...

But we remember that one really wishes for what one doesn't have. Absolute chaos wants absolute order.

The Universe likes the beauty of perfection most of all.

It's a sum of uniqueness typical of nature from birth and perfect beauty that order creates...

Unique perfect beauty is probably the most precious thing in the whole Universe.

The unique tends to become the perfect...

Rapidly spreading to all directions, the inkblot of the Universe which is full of uniqueness and chaos, above anything else craves to become a perfect immaculate dot back again.

Perhaps, some day in the future it will really happen. The universe will shrink back into a dot and will gain its original perfection. But perfection cannot be stable in this Universe, and that's why as soon as the Universe becomes a perfect dot, it will soon be reborn and it will cause a new Big Bang. And the chaos will start again.

2268. About mystical nature of things.

Unique and perfect things bring luck and energy. As if they are warm and living so that one may feel the energy of the heart of the Universe in them. As if the energy of the whole Universe is concentrated in a single spot. And this spot is now in your hands.

2394. Murphy's fifth law implementation in dynamical astronomy.

The Universe craves for chaos (the inkblot concept). The forces of order are aimed at creating perfection and some structure. Hence, there are two types of forces which are opposed to each other.

When it comes to human life, the force of order is represented by human beings (while wild nature has its own laws).

Thus, any unruly thing that a person leaves unattended, will definitely go wrong: slowly at first, gradually gaining speed until it gains critical mass after which there will be a blow and the whole thing will turn into chaos.

2405. The power of beauty.

Beauty brings joy. The joy turns into pleasure and pleasure turns into happiness.

And beauty brings good luck and draws money as well. Beauty gives health and prolongs the youth. Beauty is great power. In fact, our whole Universe exists because of beauty. At least, that's how it appears to be to onlookers.

3006. Perpetual motion machine.

Order tends to organise chaos and chain it.

But chaos craves for freedom. And freedom is a sensible awareness for chaos. This is that very perpetual motion that makes the sun shine. Chaos is endlessly various as it's a god of novelty. It figures out many ways to escape order. While order is aimed at perfection. Chains, cages and walls constructed by order become more and more perfect.

I can even imagine that sly order uses the desire of chaos for escape as a hamster in a wheel to spin it and produce electricity.

In the great scheme of the Universe we see two things. On the one hand, stars are formed while on the other hand, they are destroyed, but... it's not that important. But the most important thing is the very black inkblot of the Universe, that very chaos that expands endlessly. It never stops. The Universe expands and its edges are stormy with chaos while the centre is filled with old galaxies full of stars. We may only suppose that Order is some kind of virus that needs chaos to feed on it and use it as a material of construction. Perhaps, Order itself can get its GOAL only after having got the chaos DNA that has novelty and energy.

So who will win in this struggle? - Nobody. The essence of this struggle is in the moving forward and if something wins, the moving will stop and it will be the end of the world.

3111. It's better to be in joy than in sorrow.

The only difference of a positive-minded Universe from a negative one is the good mood of its inhabitants.

3195. About needs of energy.

Order and perfection are materials. Energy is something that creates material out of chaos. And energy is necessary for destroying material and releasing energy.

It can be said that there exists only the energy of the Universe, the energy of chaos. And this energy aims at order. It aims at perfection, to become material and structured by itself. Thus, when we say that money aims at being transformed into something perfect, we can conclude that this free energy looks for tools that would transform it into something more perfect.

However, it should be remembered that the direction of energy is a two-way traffic, it aims at both the structure of order and the chaos of freedom.

3.420.

The faster we move through space, the slower we move through time. What does that say? - That space and time are two interrelated entities, obviously two sides of the same coin.

"What does speed have to do with it?" - Obviously, accelerating, we begin to catch up with time, equalizing our speeds.

"What does that give us?" – First, it gives us an understanding that time is a state of motion of the flow of energy. The material is that which has lagged behind, that which has slowed down, perhaps has died. Things die when they stop and time overtakes them forever. Secondly, the speed of galaxies, stars and planets, forming a single vector, creates a local time constant, which, obviously, in different galaxies will be different.

- Does this conclusion have any practical value? - Speed up and you'll outrun other people's time. The greater the order, the greater the speed. Things die when they stop. Light carries mass. What does that mean?" This means that the slower you move, the more energy is taken away from you, the faster, the more energy you have left and even added.

3.429.

There's a huge correlation between rotation and time. The rotation implements the wave model phase and protivogaz. Rotating object that accelerates time, accumulating his energy, then slows down, transforming this energy into other forms

3.431.

Rotational motion with the flow of time allows you to accelerate

in the creation, and braking against the speed of time destroys the created and allows you to release energy to let it into new creation. It is necessary to build in one cycle, to destroy in another.
3.438.
The lower the speed, the faster time flows, that is, time increases. Deceleration increases, not decreases, the Delta grows. Light is something that dies instantly and is born instantly, that is the nature of waves. Those who have slowed down live longer, but their death is no less prolonged.
3.439.
Approaching a living object to the speed of light will not so much create a temporary anomaly as destroy the object. The cycle of life and death will be shortened. Living matter cannot travel at the speed of light, but information can.
3.449.
The speed of light is great, but the speed of darkness is greater, the speed of darkness is absolute. The darkness this is the time.
3.457.
The system does not slip into chaos but grows towards it. Everything grows. Order is the containment and the degradation of chaos whose aim is to prevent it from growing and to start bearing fruit. Fruit bearing is the goal of order. By hindering growth, you create a better product or growth in the right direction.
4160.
Solidity is an illusion. Anything material consists of atoms the distance between which is larger than their size. Thus, the whole Universe can be called solid as distances between stars resemble the ones between atoms.
4164. 7 questions of string theory.
If we return to wave theory and string theory, can we say that parallel universes exist in the same space with the relative displacement of the basic wave function relative to each other?
Can we say that the time shift between universes is a clock shift between the parallel processes that form the information flow of each individual Universe?
Can we consider black holes artifacts of space with an abnormal

wave offset, a bit of a strange Universe?

In the framework of the information model of the Universe, is it possible to compare this functional wave displacement as a sample of the existence of a parallel process-like those that exist in computer systems? Is the parallel universe nothing more than a parallel information process?

Can we say that the basic constants - such as time, the speed of light, etc. - in a parallel Universe are shifted relative to our Universe?

Is it possible to assume that the big universe is a kind of processor operating at a near-light frequency, where each nested parallel universe exists in its own temporal stream displacement?

Is it possible to compare parallel universes with radio or TV channels, where each frequency or shift has its own information flow?

4165.

For the next hundreds of years mankind is going to create virtual worlds, while achieving the concept of nesting. Every Universe tends to create a thousand new universes within itself.

4198. Inevitability.

In essence, Drake equation that estimates the number of extraterrestrial civilizations, is a right thing. However, the problem is that any civilization that achieves a certain progress, soon becomes doomed to death.

The transition into digital state of existence. The evolution of a human being into a stone is the latter end of any intelligent life. They simply pass from the material world into some higher digital level, or decode the digital code of the Universe and connect to it in order to dissolve in it. That's why mankind is unable to find any evident signs of other intelligent civilizations.

4.232.

A grain of sand is the most valuable thing in the universe.. The simpler and smaller a thing is, the more huge and valuable it is. The bigger a thing is, the smaller it is. Big things are definitive, small things are universal.

4249. Heads or tails.

In a parallel universe things may be the other way round.

4366. Gravitational plasma engine. GSU (gravity propulsion).

An electric motor in space must create a gravitational field and operate without converting electrical energy into kinetic energy (jet).

A spacecraft equipped with a plasma power plant that converts hydrogen into electricity, and then creates a directed gravitational field, interacting with the gravitational fields in the body of the galaxy, will be able to travel interstellar distances in a reasonable time.

Such a spaceship could move in the space of the Universe like the stars and galaxies themselves. Or, for example, by creating a directed gravitational field, repel other space objects that have gravity.

4.606.

Simple things are per se. These things are like a box where the universe is hidden. A little thing on the outside is huge and versatile inside. A little thing outside and thousands of things inside. Besides, this thing is changing its form constantly as various essences of it elevate.

4.639.

I don't like Gods because gods are idols. The real God alone is our reality, our world, our universe. I don't like what you call heaven, because heaven is hell. Hell and heaven are one and the same. The real Paradise is our real life, which is neither heaven nor hell, but just life. I don't like good, because what you call good and evil are one and the same, and good intentions pave the way to hell. But I'll tell you more, heaven is the front of hell, its front parlor.

4.860.

From the perspective of the big universe there is no difference between ant and an anthill, a tree and a forest, a king and the beggar... It all is absolutely negligible.

4.891.

The arrival of Maitreya is characterized by the following natural phenomenon: the oceans will shrink in size so that Maitreya can easily cross them. In addition, the true Dharma will be revealed to people so that a new world can be created.

Soloinc has been to Peru, Mexico, USA, India, Sri Lanka, Japan, China, United Arab Emirates, Europe, Egypt, Israel, Greece, Easter Island... There are about 80 countries in total. The oceans have become small.

The Internet, through which the philosophy of Syntalism is spread, has connected the truth with each person's phone and computer, reducing the world to the size of a dot. A huge point, as big as the entire universe.

5.316.

The greedy go to hell. The passionate know no bounds, their sufferings are infinite, just as the universe and time are infinite.

5.561.

Syntalism is a peer-to-peer religion where everyone can become an Angel, equal among equals. There are no seniors, no juniors. A small fire is equal to a large fire, and size does not change the nature of things. Everyone who is called can, by choosing himself, become the chosen one, and the chosen one is a star, a huge star in the vast boundless universe of darkness and billions of other stars. There is enough space for everyone, the darkness is boundless. After becoming a star, a person gets enough energy to act in the real world and realize all their dreams.

5.597.

An introvert is someone opposite to a balloon, someone bigger on the inside than on the outside. On the outside, the introvert is like a dot, and inside there is a whole universe of fictions and fantasies.

5.701. Pulsing universe.

There is a star - it's a white hole and there is the opposite of it - it's a black hole. Through white holes energy comes to our universe, through black it is utilized, keeping the balance of energy. White holes are responsible for the expansion of the universe, black holes are responsible for compression. The older the universe, the less white holes it is, and the more black holes it is. Perhaps, there are two universes. They are like hourglass smoothly folded to a point, then spread out. From one energy flows into another and back again.

5.703. A model of the rebirth of the universe.

The universe is constantly transforming from white form to black form. It will be filled with stars (white holes) and black holes.

5.705. Universe Yin and Yang.

Expansion of the universe is an optical illusion, rather, there is a kind of vortex hydro-press, when light penetrates into darkness and darkness into light. The expansion of white holes (stars) into darkness is accompanied by the expansion of black holes into light. Essentially, a black hole is a star in the space of black matter, and a white hole is a star in the space of white matter. And the system is closed like a ball of Yin and Yang.

5.706. Energy balance.

We can assume that white holes (stars) and black holes are two mutually degenerating, related entities, on the one hand, it is a black hole, on the other - white.

5.707. The emergence of the universe

The universe expands outward, absorbing darkness, and inside it black holes expand, absorbing the matter of light. At one fine moment there is a process of confluence of black holes and a huge central black hole is formed, the mass of which is so colossal that the black hole collapses under its own weight and a new universe is formed.

5.708.

We believe that the human brain is the most advanced neural network and artificial intelligence system in the universe, and it should be developed. The development of neural systems and artificial intelligence creates unhealthy competition for people, against which the weak degrades and dies, creating a threat to the existence of the human mind and society. In fact, the development of artificial neural systems reminds us of the development of black holes, pulling the material universe into the imaginary universe. Man creates a virtual universe, and the virtual universe absorbs man. The growth of matter causes its symmetrical absorption of nothing.

5.709. The imaginary universe and black planet.

We can say that like positive and imaginary numbers there is our universe and imaginary universe, where black holes and black matter are the stars and space of the imaginary universe. I will assume that black holes also have black planets around them like those that have white holes (stars) around them. Different types of stars like white and red dwarfs are transitional forms between white and black holes.

5.714.

The harmony of black and white holes in the universe in an example of generative design.

5.773.

Money in the universe as dirt. Money is the product of the transformation of energy from one form to another through movement. The initial forms of energy are time, hydrogen and water. As you know, there's a lot of that stuff in the universe, so if you're short of money, it's probably not about money.

5.774.

The sun is an energy extreme. Water is the harmony of energy, and a black hole is the extreme of the energy of absence.

5.812. The death of God is the creation of the world.

You know, death, too, can be love, death beautiful and attractive. Because death is always a harbinger of new life, dying, we generate new. When we die, we create. Creativity and joyful, so joyful death. The man who overcomes the fear of death becomes a Creator, dying, he creates a new beauty and therefore happiness overflows his soul. You asked me if God was alive? No, God is dead. In creating our universe, God died, reborn into his creation. God is everything now. Our whole universe and even you are a part of the dead God, his little seed, which can also grow and, dying, create its own world.

5.911.

An angel is a person whose spirit is formed by generative philosophy. It is a very lightweight and at the same time strong composite construction, used in the aerospace industry. In order for a man to enter the universe and conquer space, he must become an angel.

5.918.

How "Variothoughts" works. First they drill your brain without anaesthetic, then they put the seed material in it (they say it is the seeds of truth), which causes your brain to explode and in this unbearable pain there is a flash of light and an explosion. And everything. For the next 13 billion years, this universe of mind slowly grows, eating the last remnants of the darkness of stupidity in your head.

5.975.

Given that time is motion and motion-it seems to have something to do with the motion of our sun, galaxy and the universe as a whole. It is known that in some galaxies stars move clockwise, in others - against. We can assume that time in these galaxies goes in different directions.

6.593.

At the lowest level, the most elementary particles that make up the universe are Nothing. Nothing, grouped in the System, forms an elementary essence.

6.594.

The point is the pattern of restraint of Nothing. Holding back its energy, Nothing reaches a critical point and explodes, so there is a new universe.

6.599.

God is information about all unique perfect forms created in a given universe. In fact, our universe was created by God to generate new perfect forms.

6.602.

Truth is what gives form to energy. God shaped energy and so created our universe out of pure energy.

6.603.

The creation of the universe was not an explosion from a point, but simply from an elementary point, the formation and structuring of the energy field began. The primary form of energy was time, then time split into light and then light, already changing its form, began to take other forms.

6.684.

Humility allows you to stop resisting reality and let in all the energy of heaven and earth, becoming a conductor of the energy of the universe here on earth.

6.732.

Once in a dream I saw a world where balls of liquid time floated in the space of non-existence and one of such spherical formations filled with time was our universe. The universe is a ball filled with liquid time, in the very center of which a small dot floats. The very point within which all our illusions live.

6.869.

Evil does not exist. There is only Nothing that loves Beauty. Good is beauty. And there is a process of growth, when like the universe from a point, from nothing begins to grow good. While beauty is small and weak, it is called "evil", but when it grows stronger and gains strength, it begins to be called "good".

7.146.

An objective universe originated from a subjective point.

7.156. Escape from the wheel.

By pushing the boundaries of reason, man escapes his ignorance. Initially inside one ignorant and sits at the point around which the universe of knowledge, unavailable to this man. To gain knowledge means to make a big Bang, to push the boundaries of the mind and from the point to become a reasonable person.

7.356.

The Christian cross is the equivalent of the mathematical symbol X. X is everything. X is the greatest and the smallest number in the universe and the matrix of all possible states.

8.144.

Universe consists of zeroes and ones, if you are one, you don't care about zero being nothing by its own, a one makes ten out of a zero.

8.848. Multidimensional Universe.

There is nothing unambiguous in this ambiguous world.

8.918.

A spatial hole-digger is a thing that digs holes in space.

8.943. A snowflake.

The expansion of space-time continuum resembles the expansion of a snowflake. Every moment is full of the creation of billions of variants of the Universe. Every moment you simultaneously stay in all variants of events. The variant you believe in most of all, makes your present reality.

But in essence, you can believe in any variant and thus, turn it into reality, it only depends on the force of belief and the ability to concentrate. In fact, something that you expect and consider the inevitable, becomes created in reality. Your brain considers that very reality which you expect to see.

8.956. Qubit.

The analysis of superposition in quantum mechanics proves the basic philosophic points about the existence of good and evil, yin and yang, darkness and light as well as points about negative and positive perception of the world. The world simultaneously stays in its two basic states. The smallest brick of this universe - the Qubit- is simultaneously both 0 and 1.

8.994.

Can stars and hot planetary cores be called a special case of quantum computer systems? A more general case of a quantum system is galaxies and the Universe itself.

8.995. Primary point.

Being everywhere, always and in all its conditions at the same time, point is like Universe. People and scenarios are very similar, with many conventional models. Library of models is no as big as it seems at first sight.

9.263. Space is environment for energy transmission.

Cryogenic temperatures reduce resistance to zero and the phenomenon of superfluidity emerges. Under such conditions losses at energy transmission disappear and its transmission for huge. unlimited distances becomes possible. I'd like to note that it's very cold in space and in particular, in the Universe. On superfluidity I'd like to note that time, light and the space itself exhibit the characteristics of liquid.

9.264. Flowing Universe.

Given the phenomenon of superfluidity under low temperatures and the fact that energy is very similar to liquid, it can be assumed that the Universe not only spreads in all directions as a result of the blast, but it also flows around and above due to the lack of the emptiness resistance.

9.351.

Black holes are energy channels with ultra-high pressure and low temperature, having superconductivity and designed to transmit energy, including in such States as matter and space.

It is believed that the peculiarity of black holes is their opacity for time and light. Time and light cannot penetrate black holes.

9.353.

The black hole is the very point, analogous of the one out of which our Universe emerged. It's a super-conductive energetic channel, puncture into the space of Nothing, where energy of our Universe outpours, creating a new Universe there.

9.354. Creating a new Universe.

The Universe creation process isn't a blast from a point, but the black hole-like puncture out of neighboring Universe through which energy from the neighboring Universe starts to outpour into ours, filling its emptiness. Basically, it's like a start of a new child process, linked with the parental one.

9.356.

The amount of energy in the Universe is endless.

9.357. Linked vessels system.

The space of universes is the structure, consisting of countless bubbles of the universes, that are interlinked by energetic channels of the black holes

9.358. Information virus and running energy.

Information is the virus trying to structure and shape energy. Energy is chaos and content, information is order and form.

Moreover, energy kind of tries to run away, flowing into space of Nothing through the black holes and creating new Universes, but information chases and structures, trying to shape it.

9.453. Flat universe.

A plane is a plane no matter how it's being wrinkled or rolled. If a plane is formed into cube and pressed under high pressure, would this turn it into a cube? Evidently not. Our Universe is a particular case of such a plane

9.461.

There is no constant coordinate system in this universe, space is relative here and is in permanent motion. Coordinates here are not absolute, but relative to other large objects.

9.484. DNA of the Universe.

A point is the analogue of a living cell, it has analogues of DNA and RNA. That is, any point has all the necessary information about its structure. Any point has enough information to become the Universe. A point can be decomposed into any figure; any figure can be reduced to a point.

9.508.

Vacuum and emptiness should be created around a point to let it get a chance to become the Universe, vacuum will suck energy out of it and will break the hole into the parallel Universe....

9.510.

One of the purposes of the black hole is recycling of all the matter forms and turning them into their original forms: oxygen, deuterium, lithium and helium.

9.538. All objects are the same, except superconductors.

The smaller the object is, the more energy it has. The bigger the object is, the less energy it has. The constant process of energy conversion from one type to another constantly occurs inside objects. For example, energy conversion into the matter. But the point that is the universe doesn't use its energy for it, this point is a superconductor.

9.540. Point aspiration to become the universe.

The most sustainable and stable structure is the system with the least energy. The more energy a system has, the more unstable it is and the more it aspires to decay.

9.551. Quantum space.

In the world of atoms and electrons, there is no movement, and, consequently, there's no space and speed. Electrons do not move anywhere, but exist at certain points (states) with one or another probability. In fact, if there is no speed and motion, it may be possible to move this way in our universe, too, simply changing the probability of finding the object at a particular point.

In fact, we can say that there's no space at the electronic level, but there is only the density of electrons. This system is somewhat similar to flash memory in modern computers. The quantum space is very similar to the information space – the system's memory.

9.563. Point.

The fifth dimension is some coordinate relative to current coordinates of our Universe. It can be assumed, that quantum space is five-dimensional and that's why its laws are so drastically different from laws of four-dimensional space of our macrocosm.

The phenomena of five-dimensional space explain quantum teleportation and probabilistic nature of electron coordinates. But this space is eight -dimensional, not five -dimensional, so all our four-dimensional space is nothing more than point within eight-dimensional space.

9.564. Theory of being No 7.

In the black hole the Universe can stretch into an infinitely thin string. However, it should be added than this string is the super-conductor through which endless amount of energy can be transferred, and its infinite thinness, relatively speaking, can puncture three-dimensional space and create new universe on its other side in eight-dimensional space. Let me remind you that that in eight-dimensional space our four-dimensional space is only a point.

9.569.

There is no distance inside a point, as distance is the straight line between two points. Since there is no distance, there is no motion, no time, no acceleration. In other words if one gets from eight-dimensional universe into our fourth-dimensional one, he will be able to get to any of its time and space.

9.570.

The big band is some kind of illusion. Our universe was and continues to be a point. A four-dimensional point in eight-dimensional space. Basically, the big bang is the relative process of eight-dimensional space transformation into four-dimensional one, resulting in the release of energy.

9.573.

Every point in eight-dimensional space is four-dimensional universe. Every point in four-dimensional universe is two-dimensional universe. Basically, every point is essence, that is a point on one side and multidimensional universe on the other side.

However, given that time is zero, in reality these universes are one-, three- and seven-dimensional + 0 (time).

9.621.

A human being is a special case of a multidimensional system, externally resembling a dot while internally being a whole universe...

In the internal universe there are inexhaustible stores of energy, if one gets out of the dot it becomes possible to produce a big bang and create a new universe already in the real world.

In order to make a dot a universe, a dot has to become a superconductor providing energy from the inner world into the outer world. In order to make a dot a superconductor it takes extreme conditions and Nothing. It's Order that creates extreme conditions best of all, while Freedom is Nothing and Void, a space for new life.

10.259.

The mass of a black hole is such that all movement within it disappears, creating a space-time anomaly. It is the black hole that generates the point in the distant past from which the universe originated, concentrating energy into a single point where there is no time or space.

10.504.

What is the flapping of a butterfly's wing that easily changes the whole world? The life of one person in the history of mankind. The existence of all mankind within the history of the universe.

1097.2. The structure of the world.

- The world is a giant energy field with information encoded in it. The entire paradigm of this information's existence is based on the fight for energy. The more perfect and unique this information is, the more energy it can obtain. Matter, space and time do not exist. The world is an information system existing in an energy field.

- The Universe is a huge self-learning information system whose very raison d'être is to create new information featuring a certain level of uniqueness and perfection. All the mechanisms and principles of this system are aimed at achieving these goals.

- In this world, everything is information – matter, time, space, man, soul, thoughts, ideas, and so on.

-Death and destruction do not exist. Everything that was once created remains in the information field forever.

- Time, space and matter are the information about the coordinate system and the structure of an information entity.

- The driving forces behind the development and growth of information are six endless aspirations that have been accomplished in the mechanics of the Universe: to exist, to create, to destroy, uniqueness, perfection and expansion. They are all in a state of balance with each other and form the seventh control function, or Harmony.

- Man is one of the Universe's systems designed for creating information.

- The human soul is a small Universe.

- A man completes a number of missions throughout his life: the Creator, the Cleaner, the Companion…

- A man's goals that are subject to nature's laws determine his destiny and the amount of energy that the man will obtain to realize it. To manage energy, the organism needs a certain level of perfection and uniqueness. If it is not sufficient, the man's destiny will use various tests and difficulties to train him. A major goal requires a lot of energy. The man must be ready for it. Defeats build character and lead to perfection. Victories give the man more energy to take a step ahead.

- A woman personifies perfection and a man personifies uniqueness. Both of these origins are present in each person in varying degrees.

- Perfection wins energy back and uniqueness obtains it due to its novelty.

- Chaos engenders maximum novelty and uniqueness. Order engenders perfection.

- The main meaning and goal of human life is to participate in processes related to creation and perfecting. This gives energy and strength to a man. A man is divided into the external and the internal. Even in extreme hardship, there is always room for the internal physical, spiritual and intellectual growth. The first thing a man should create in his life is himself...

- The processes of creation and destruction are closely related. Energy and location are required for creation. Processes related to destruction release energy and make room for new challenges and goals.

3.1004.

Curiously, a 20% chance of winning versus 80% of losing can easily win if the time and place are right. There are always four of them... he, maps, time and place... All this makes it very difficult to understand the theory of probability. Volume probabilistic model.

3.1473.

The rotation of the point generates the universe. A fixed point is nothing, but if a point starts spinning it will turn into the universe. The squirrel in the wheel has a great chance of immortality.

3.1513.

Zero is the point from which our universe was created.

3.1972. Point and universe.

Truth is a scalable entity that can come from diversity to unity and from unity again to diversity.

3.2076. All roads lead to Rome.

Indeed, the greater the knowledge, the greater the ignorance, but... a point becomes a universe, and the universe becomes a point. Having reached a critical mass, the manifold again adds up to a point. All roads not only leave Rome, but also lead to Rome.

3.2077. A point is a beginning and an end.

They say the more you know, the more you don't. This is true until critical mass is reached, then the carriage turns back into a pumpkin, and the universe into a dot.

3.2330.

The universe is expanding with acceleration for the same reason that winds blow into a low-pressure zone. Nothing and the void draw matter.

3.2331.

If we compare the expansion of the universe with the wind in the atmosphere, then there must be a reverse circular flow of compression.

3.2370. Prism effect.

The paradox of space is that the process of expansion of the universe is at the same time its process of suction into a black hole. That is, the universe simultaneously expands and develops.

3.2379.

The expansion of the universe happens with acceleration and the tightening of matter into a black hole the same happens with acceleration. Coincidence? "I don't think so."

3.2386.

A black hole is the universe. The universe is a point in a black hole. In fact, this black hole itself is this point.

3.2393.

Expanding, you squeeze. Disconnect, connect... The universe, expanding with acceleration, with the same acceleration adds up to a point.

3.2403.

When a black hole reaches critical mass, it inevitably transforms into a new universe. – Why we don't observe this process? – Too little time has passed since the beginning of our universe, they are still small. Over time, the fusion of black holes will create one mega-huge black hole that will consume our entire universe, turning it back into a point that, reaching a critical mass, will explode again and a new universe will emerge.

3.2916.

The meaning of creating the universe is to create God. There is no

God yet, but he creates himself from the future with our hands.
3.3124.
The force of gravity depends on the time and, therefore, the speed of the object. So there is a relationship between mass and time. The greater the mass, the slower the time?
4.1732.
The feeling of power and power is associated with getting rid of duality. If you get rid of the I and Mine, you will become one with God, you will become God yourself. All the colossal power of the universe will become you and yours. All the people, all the money in the world, all the living and inanimate will become you and a part of you. You are chaos and order, system and chance. You are everything. Your strength and energy are infinite. Your power is immeasurable.
4.1736. The egg, not an ordinary one, but a golden one.
The death of Koshei is hidden in an egg. Tale "Riaba the Hen" An egg is a symbol of fragility of illusions. If an egg breaks, a disaster will occur. An egg is a crystal sphere of illusions, a universe where a person lives. If one breaks illusions, it will be a disaster and the world will perish. The evil, the anger, the hatred will break free, if one opens Pandora's box. An egg is a symbol of Pandora s box sealed permanently, a symbol of genie in a bottle, symbol of a hamster in a wheel An egg is a permanently sealed golden casket, which will set the evil free and able to start making problems, if opened. When illusion are inside one's head it's another story, but if the thing inside get into reality it won't be pretty.
4.1850.
Christians say that God is a dove, Muslims call him an eagle, the Jewish name him a sparrow and for Buddhists he is a dodo... Other religions and sects give him the names of other bird species. As for Syntalism, it considers that, in a broad sense, God is a bird and, in a narrow sense, he consists of body cells. If we expand the understanding of God even more, God is the universe and God is electrons, atoms and quarks. God is everything and God is nothing.
4199.4. Stones.

Technological development will bring intelligent life to a digital and energetic level, and it will either create new levels above our Universe, let's call them "new parallel worlds", or it will join the already existing systems that were once created by other civilizations.

4.2078. Those living in illusions.

You will never attain integrity if you keep rejecting extremes. The boundaries of shapes go along the extremes. By negating that, you will inevitably turn, first, into a stub and, then, into a point. To be means to have a shape. In order to be, you should be resigned and combine extremes in yourself. The three-dimensional universe has three vectors of the extreme. You will be impossible if you choose only one or two of them. By rejecting this or that extreme, you will be false and illusionary like a drawing drawn on a piece of paper. Like the midsummer night's dream. Like the last agony of a dying person.

4.2097.

The hardest thing is to be consistent in inconsistency. In the end, you become so predictable that there is no question of any accident. I'll tell you the truth, man is incapable of accident. Chance in this universe is owned only by God.

4.2122.

Indeed, the black hole is the backwards Sun. A star doesn't let any energy out. A black hole consumes energy. A star is somebody who loves. A black hole demands love.

4.2202. Black is the color of emptiness

Things consume light. Color is a free light that wasn't consumed. All things have colors. Colors that weren't consumed by them. Unconsumed color is what they already have. Black is the color of greed...when everything is consumed because it's empty inside. Inside is the black hole.

4.2332.

The point is that even though the devil is the hell's king...the true God of the Universe is truth and beauty. God is law and the sun is a slave of the law just as any other sinner. For God both the Sun and a tiny ant are all dust...

4.2333.

The Sun is an idol. God is the Universe. The Moon is a fake idol. Venus, Mars, Mercury and all the other people are hell, just as the Earth, but worse.

4.2423.

Planets are the same stars, only immature, whose growth was stopped by the shadow of the star, which, using its power, took away all energy resources. But the planets are not offended, the hell that reigns in the Sun has little to do with joy.

4.2489. Man is soul?

Father, son, and Holy spirit is a metaphor where the father is the sun, the son is man, and the Holy spirit is time and the universe. The Holy spirit is being.

4.2492. Quantum slip.

Everything is possible in this world. Trillions of quantum universes are created at every moment. You are the creator of the universe. Every movement you make creates a new universe.

4.2567.

The development of thought went wrong. The Greeks prayed to Zeus, the personification of the Sun. The Egyptians also considered RA. Later Egyptian thought came to Amon RA. Christianity, having analyzed this concept, came to the conclusion that the worship of one Sun is idolatry and is a higher entity that is All. God is the universe. God is being. There are trillions of idols like the Sun in the universe. Plus Christianity pointed to the sacred essence of man himself, proclaiming his soul the son of God, part of God.

4.2658.

It is true that our world is ruled by the devil, but our universe is ruled by God and his angels.

4.2684.

The squirrel that runs in the wheel is a resident of a two-dimensional ball in a three-dimensional universe. Wherever she runs, she always runs back. The idea that one could choose to go up or down by starting to grow does not fit into her two-dimensional consciousness.

4.2761. Three-dimensional point.

Space and time are forms of space created by the ideas of the mind. If it were not for consciousness, the past, present, and future would exist simultaneously in one instant of time. However, you already know that the universe did not originate from a point, the universe is a point.

4.2811. The universe.

To him who wakes in the night, it seems that the Moon is God, and the stars are her servants. He who wakes up in the afternoon idols the Sun that feeds him. However, those who are smarter, know that many lights in the sky are huge galaxies and Suns, like the idol of the day. The sun is an ordinary star, which does not stand out from a trillion similar grains of sand on the ocean floor. By the way, the ocean is more like God than the Sun.

4.2821.

Curiously, black is not just nothing, but the absence of waves in the visible range of white. That is, the waves there may well be, just in other ranges.

4.2824.

Interestingly, the visible white light takes only 1% of the spectrum. What arrogance are you talking about if the black is 100 times bigger than the white? The white is like a spoon of honey in a barrel of tar – something tiny but important. What's more, the white is not necessarily stronger than the black. Waves in the ultraviolet, X-ray and gamma ray ranges carry much more energy. The violet and blue colors have the most energy in the white color range.

4.2826.

Black-black discord. Black this as a factor X. For the black can hide anything, and as black can hide behind anything. For example, the Sun, which you see yellow or white, the greatest flow of energy transmits in ultraviolet, which you perceive as black, and the smallest in infrared, which also seems to you black.

4.2838.

In this material world, everything is matter, that is, energy, which has been given one form or another. Intangible inside this universe does not exist. Matter this there is information in unity with energy, but to sever this unity impossible.

4.3392.

There are two universes: one is inside the point, the other is outside it. In one universe the dot is God, in another Nothing.

4.3481.

The Yin and Yang symbol is false rather than true. He says white and black in half, but that's a lie. The Golden mean is at 0.618. But I would also advise you to look at the distribution of matter and its absence in the universe, in this case, you will notice that black and white are like bread and salt, where white is not more than 1%.

4.3937. Black and white.

Black is the biggest color of all, it is 100 times bigger than white and 700 times bigger than any other pure color. White is the color of modesty, black is the color of strength. The weakest color is red, green and yellow. Blue's a little stronger. White is proud, it is seven times stronger than other colors, but 100 times weaker than black.

4.3963.

The universe is like a sponge, which swallows up one day and compresses the other.

4.3973. In the boiler.

They say there was a big bang and now the universe flies frantically fast in different directions. It can be so, but I had a dream where God poured a glass of water and put it in a microwave. Time and black substance are boiling and bubbling.

4.3979.

The universe had not exactly exploded out of a point, it's more likely to be put in a microwave and heated strenuously.

4.3980.

It is felt that our universe is either warming up in a microwave, or they just turned the power on in a solid-state drive. Memory areas that are still free are black matter, and other areas have something recorded in them.

4.3981.

The universe reminds me of a pancake, that is being fried either in a microwave or on a pan in the butter.

4.4258.

One grain of sand is enough to create a whole world, our world was created from a point/Each grain of sand in this world is enough to build a new universe.

5.1088.

The universe is multidimensional as a tree. Virtual world of dream, virtual computer world, the world of psilobicin illusions. It is interesting, the real world is an illusion of what type? If it is known that Armageddon must destroy the illusion of lies.

5.1279.

To know the idol is to know the sun, to know the truth and harmony is to know the universe. But he who knows the idol knows perfection, for the sun is perfect. But he who knows harmony knows black holes, billions of other suns, and the system of the universe.

5.1399.

If light is accelerated by more than the speed of light, it will be transformed into matter and a new universe will be formed.

5.1418. Superluminal material and energy generator.

Exceeding the speed of light leads to a change in the energy of its shape and transformation of light into substance, releasing a huge amount of energy. There is a probability that a new universe will appear at this moment.

5.1913.

Creating a new universe does not require much energy, but requires a lot of imagination.

5.2196.

Love is attention, when you feel attention to yourself, from all points of the universe it is the love of God. God sees everything. And you, and all your thoughts in the palm of the whole universe.

5.2197.

They say Soloinc wrote "Variothoughts", this is not true. "Variothoughts" was written by the universe. Soloinc is just an avatar of God, one of the countless angels of Nothing.

5.2228. A black hole is a one-dimensional point.

A black hole is an artifact or degradation of one-dimensional space in three-dimensional space. Initially, the whole space was one-dimensional, then its development created an explosion of the three-dimensional universe. As the explosion weakens, the three-dimensionality disappears again, returning to the one-dimensionality of black matter.

5.2352. The abyss of suffering.

If love is not fed, it will die. If beauty is not fed, it will die... When beauty and love die, there will be a black hole and all the guilty and innocent in the radius of this disaster will also perish to hell...

5.2947. Praise the genius.

Awards and prizes are simply necessary for a genius at the end of a great job, so as not to go crazy from depression. When you finish a huge job, it's like you die and fall into the abyss of emptiness. The universe of emptiness and loneliness is endlessly maddening. Here, of course, external moral support would come in very handy. Finishing one project, genius, of course, is embarking on a new, but new very small, as grain, and gap emptiness a huge as universe... Very small needs love.

5.4370. In the depths of the blue screen.

What is the world? A blank, white piece of paper, an ocean of energy. What are objects? Patterns of lines on the white piece of paper, the movement of information. Objects are an illusion created by their cognition of one another. However, a stone cannot cognate itself, so the stone does not exist. You can, though. This is why you define the stone instead of its defining you.

5.4426.

The Universe is not a point of light in the ocean of darkness – it is a black point on a white piece of paper... What you see is something negative.

5.4440. About opportunity.

Humility is the absolute idea that the absolute does not exist. Humility is the idea of a God who does not exist. The universe is what is and what is not. God is what is and what is not. Truth is what is and what is not. Objects exist and do not exist at the same time. Humility is acceptance of Einstein's theory of relativity and quantum logic. It does not matter whether there is an object or not, the main thing is that it may or may not be here.

5.4484.

In unity there is no harmony, unity, finding a form that gives rise to freedom. Freedom is sand. Freedom is water. Freedom is poison... When these three freedoms become one, they become the concrete of the material world. This world is made up of sand, love and vices. If you remove the senses, it will turn the world into dust, depriving it of material form. They say that every tiny speck of this dust is the very point from which our and all similar universes were.

5.4485.

When you break the boundaries of your Self, you turn into nothing and become dust. A tiny grain of sand, the one inside of which is the whole universe.

5.4493.

In the diamond sutra it is said about the structure of "thousand thousand worlds", it is a galaxy. There are three thousand such galaxies in the universe, and this universe itself is a speck of dust... There is a point from which our universe was created (or not created, for every grain of sand is the universe).

5.5152.

It's just a moment now, but the truth is, it's the only thing that exists. All we have is one moment and nothing more. But this is the material, and the immaterial, that is, illusions, we have the whole universe.

5.5175.

The idea that you need a lot is deceptive, in fact, the universe is hidden in a grain of sand.

5.5233. Figure in fog.

Awareness comes from two sides: on the one hand, it is concentration on the tree, and on the other, dispersal on the forest. The simultaneous hardness of small and large dispersed. Big, it's like fog, like water, like air. Small as stone, small as rock. The small is the planet Earth or even the Sun, and the big is the universe. Life is a transitional entity between the Earth and the cosmos, the bio-sphere of the Earth.

5.5425. 47 words.

In "Variothoughts" there are about 47 thousand texts, the meaning of which is to reveal the meaning of very simple words and concepts. Simple words are as vast as the universe and have many meanings within themselves. The meaning of "Variothoughts" is to reveal the meaning of words such as God, time, love, ritual, faith, hope, dream, good, justice, harmony, peace, soul, mind, feelings, beauty, truth, order, law, dream, plan, habit, passion, lies, wisdom, stupidity, fear, Vice, devil, demon, despondency, greed, lust, laziness, power, strength, chance, miracle, life, death, light, darkness, disease, whip, gingerbread, health, man, Grace, luck, virtue, morality, ethics, morality, synergy, pride, animal, envy, the chosen, saints, saints, freedom, etc.

6.2018.

When you do the right thing, all the time and love of the universe is in your hands.

6.2020. Time is God.

Time is the flow of energy, the flow of hardness of matter, the flow of force. In a four-dimensional universe, time is the wholeness of space and matter, the wholeness of the three extremes. To understand God means to know time. When we say that God exists outside the time and space of being, it means that God himself is time, and time itself is the flow of space. God is a river of time, water, in which it is the matter of space, and the current is the movement of time.

6.2076.

Understand me correctly, I do not believe in green men and reptiloids, but I am absolutely sure that examples of reasonable life in the universe are more than enough. I don't think God is anything like or different from me, but I'm sure that God is everything and it all exists. I believe that God is the law and order, and that this law is understandable to me.

6.2163.

In fact, time is movement, movement itself is energy. After the big explosion there was a release of energy at great speed. As the universe cools down and slows down, so does time.

6.2164.

We can assume that the black hole is the effect of time turbulence, looped time, time outside of our time.

6.2167.

Our universe consists of energy and information, and time is the simplest form of energy.

6.2278.

Hardness of hard strife. The distances between atoms are so vast that it is unlikely that a solid can be considered more solid than our universe.

6.2363. Points.

The book "Variothoughts" begins with a dot. Just as the universe began with a point, so knowledge begins with ignorance. You have to put an end to everything that was "before" and start all over again. As long as you are filled with the old knowledge, the new knowledge will not find a place in you.

6.4301.

Light is the fastest thing in this universe. The more light you have, the faster you move and grow. Light allows you to fly, light is fire and energy.

6.5767. Loneliness.

At the end of year 12, I met God. I was in hell and I was in heaven. Hell was full of pleasure, but after thousands of years, these pleasures have become so painful that the strength to endure was no longer and cunning I escaped from there. Then I visited Paradise and the Union of love with God and his purposes filled my soul with joy. Thousands of years of common cause passed as one second, the creation of peace and beauty are truly joyful. Then I visited nothing, and his loneliness filled my soul with compassion as great as the whole universe.

6.5784. Waterfalls of time.

Our universe was created by love, concentrating all his love into a point, God created time and light.

6.5976.

When you do your work fearlessly, nothing can hold you back for long except firmness, but judge for yourself, how much firmness is there in this universe? 99% of the darkness belongs to you.

6.6228.

Biting a hole in space time requires determination.

7.1046. Evolution Of God.

It was thought, but God is also evolving. If the whole world is God, and we see that the universe is expanding, nature and man are developing ... then it turns out that our God is growing and becoming stronger and stronger.

7.1770. Full plate requires a lot of perseverance.

The development of the universe from a point of view takes place along an expanding spiral. The problem of the last mile and achieving perfection is the volume, not the number of steps. Expansion of the spiral leads to the fact that the volume of each new circle is much larger than the lower circles. Each step is very difficult to achieve.

7.2518.

A black hole is a kind of unfinished universe. At the end of the process, a new universe could arise at this point, but until it does, it is the point of tension of space where the energy is leaking. Just as a subjective world is created in the brain of a neurotic, a new universe is created at the other end of the black hole, different from our universe, but powered by its energy.

7.3087.

Having tied up like a fly in a web of illusions, a person immersed in virtual worlds will lose the means to perceive the real world. It is said that this will destroy the universe. But it seems to me that it will destroy only a man, because not the universe is an illusion of a man, and a man is an illusion of the universe. An illusion that has lost its connection with reality dies.

7.3578. Like the universe.

Willpower and repetition of certain actions create habit, habit creates character. In fact, character is man himself, that is, man is a self-creating entity that can create itself.

7.3579.

The universe is a self-creative essence that creates itself by obeying its own will. Beauty and perfection are the first aspirations of the universe.

7.3580.

The desire of the Universe for destruction is connected with its desire to live. Life is a constant renewal.

7.4135.

The laws of being are the DNA of this world, embedded in the grain from which it grows in the space of non-existence. The DNA of the world can be changed, but only by creating a new grain and a new universe.

7.5042.

God is not the one who created the universe. God is the one who destroyed the universe. God created Nothing, created a free place. And nothing is something from which the universe begins to grow.

7.5043. The universe is growing.

The essence of the unity of God and the devil lies in the idea that for creation you need nothing - free space. By destroying the universe, the devil created Nothing, God planted seed in Nothing. God sowed the field, and the Devil is the Reaper who separates the wheat from the chaff... When the harvest is harvested, the devil digs the field, thereby destroying the universe again.

7.5251.

The evolution of the games will lead to their merger into one mega game, which will be called the universe. Current version of the Universe is v.3.0

7.5253.

The devil has destroyed the universe many times, but you know that the main virtue of God is perseverance. Not discouraged, he creates it again and again.

7.5635.

By concentrating your illusions to a point, you will create the universe.

7.5699. The universe consists of what is not.

Nothing is an absolute point, a point where there is no space or time. It is from this point that our universe was created and it consists of them.

7.5701.

The universe is a wholeness made up of nothing. The universe consists of nothing, i.e. that which does not exist separately, but when it gathers into a whole, it acquires wholeness.

7.5810.

Everything is a number, we live in a virtual universe.

7.6006.

They say the fire is pure and beautiful. They say fire is perfection. But that's not quite true. Fire generates light, clean fire produces white light. But the colors are different, colors, connect, give birth to new colors. From light there are such entities as electricity, paint, TV, virtual and real worlds.

7.6007.

Was originally white color and no black. It spawned the black-and-white world you saw on black-and-white TV. The presence of light and emptiness created movement. Individually, neither light nor darkness produced movement. The movement created life. A little later, when white light split into other colors, it created a colored world. The more colors in the world, the more realistic and detailed it is.

7.6008.

The division of light into colors created the possibility of movement within perfection, without the use of black. It's called personal growth. And leads to growth diversity flowers and the formation reality, i.e. of truth.

7.6009.

Light is infinite, truth is infinite and varied as the thousands of colors and shades. Black ultimate.

7.6010.

Diversity allows you to grow not only from black to white, from nothing to perfection and back, but also to create life after achieving perfection, dividing the white light into millions of different colors and shades. Color gave perfection the opportunity to live after achieving perfection.

7.6014.

Darkness also grows and tends to light, mixing with light, it forms a semitone, Dismounting with color, becomes colored.

7.6015.

Truth is detail, it is the desire for color. Both darkness and light tend to color. The more detailed and colorful the image, the closer it is to reality and, therefore, to the truth.

7.6016.

Truth is all extremes and everything in between.

7.6018.

They say perfection is pure white light. But maybe that's just one side of perfection, the other side of perfection is black. The movement between these two entities creates life. Movement is life.

7.6019.

But the perfection of white is infinite. White is capable to perfection by self-destruction on thousands and millions of flowers. Forming, thereby, color realistic truth, which gives shape to our world. But who destroys the white light? Black, breaking white, creates color. The color white is defective, vicious white. It is the shortcomings of people that make our world so diverse and colorful.

7.6020.

Color is the division of light into parts. Darkness, destroying light, causes it to break into thousands of pieces, thereby creating millions of colors and halftones.

7.6021.

Striving for the perfection of the real world, you should not strive for white, and should not strive for black. First, pushing white and black, break them, and then, when there are thousands of other colors, give different shapes and create a new world. Make this world move, it will create life. Life is the truth.

7.6022. Celestial mechanics.

To gain access to energy, you need to create perfect things, giving them perfect forms. To do this, you need knowledge of the truth, it will allow you to create perfect ideas and will give you the power to shape energy, materializing these ideas in reality. Start by training your brain's neural network-train your mind, fill it with knowledge so it can create ideas. Train your spirit power to control and improve your body, for this is your first tool for implementing ideas. You must gain control of your body.

7.6023.

Truth is life, it is a thousand colors and details, the movement of forms and essences.

7.6024.

Reality is always trying to descend into chaos. To avoid this, white and black strive for the integrity and strengthening of forms. This process is presented to many as the pursuit of perfection. Pure colors and large forms are seen by many as a model of perfection.

7.6026. Black and white strive for perfection, and life for chaos.

Hating each other, constantly trying to divide and dominate, white and black create colors, shapes and movement, thus creating life. Life is beautiful and perfect, but it tends to chaos. Thus, black and white, fighting among themselves, tend to order, and life tends to chaos.

7.6027. Three-dimensional aspirations of being.

The desire of black and white to order (at the same time from each other), creates a black-and-white world where there are two poles of order. Such a world creates movement and life. The desire of life for chaos creates truth, that is, our three-dimensional color reality, where there are millions of colors, halftones, large and small forms.

7.6028.

Life, striving for chaos, breaks the order of black and white, thus generating itself. Mind is life, the essence, itself generating your aspiration into chaos.

7.6477.

Manada is a demon spirit, each a separate universe, living in a world of illusions.

7.6600. God is Nothing.

They say there is no God. There is only the Devil and the Devil's absence. The absence of the devil so happily and clearly that is nothing take everything for God. God is the emptiness and nothingness in which the devil lives. God is the universe in which the sun shines and hell is located.

7.7646. Close your eyes and imagine.

You are the tree whose only desire is to grow the fruit. You're a mother fascinated by the new life inside you. You were created to create and bear fruit. The energy of growth flows in your every twig and leaf. Your roots reach the very center of the earth, absorbing the energy of its red-hot heart. Your skin and leaves are eagerly drawn to the sun, absorbing the energy of the universe. Through you pulsates a huge flow of energy, designed to create beauty and perfection.

7.7653. Fear is the shortest way to hell.

The basic flaw in the Creator's greed at the time. Greed of time causes fear of losing time, and fear, as you remember, is the shortest way to hell. Remember, the Creator has eternity. All the time of the universe and everything living in it belongs to the Creator. Thus, if you are a Creator, you have enough time for everything you create, and the fear that torments you is a trick of the devil to make you spoil the mystery of creation.

7.7654.

All the time of the universe belongs to the Creator. The Creator is not late and in no hurry. Time belongs to the Creator. If you create, you get time and energy. If you do not create, you give your time and energy to others who are busy creating.

7.7655.

The power of thought is the greatest power in the universe. Our universe was created by thought. The energy of thought is boundless and inexhaustible.

8.1087.

The example of the creation bubble of the Universe out of a single point proves the hidden potential of any point... Thus, every person has a great potential for growth on condition that one manages to fulfil the given potential.

8.1091. Dangerous curiosity.

The Universe originated from a point. The energy of an elementary particle is great and if it gets released, there will be a big bang.

In the past, our Universe used to go through this.

Having exploded, the point turned into the bubble of the new Universe. And this bubble still keeps on growing. And it will keep on growing until some idiot decides to split one more point.

8.1168.

The Universe is a particular case of an information system where exponential mass accumulation takes place in.

8.1169.

Analyzing the inflationary theory of the Universe (the theory of bubbles), the concept of "Googleplex", quantum theory, Cybernetics, the theory of Syntalism and drawing conclusions from this, we can say that.

Essentially parallel worlds ..."these are the very bubbles created within our Universe. That is, the same equal universes formed by the big Bang as our own... in the neighboring address space, that is, with a certain offset from us. And at the same time, along with understanding the nature of parallel Universes, we understand that there is also a Parent universe, that is, the one inside which all this is located. And another interesting conclusion, since a quantum always has two simultaneous States... Indeed, we can say that there is a third entity in this system, such as the positive and its clone negative universe.

And you can also go deeper into our Universe, that is, go to a level below.

8.1170. 5 types of possibly existing Universes.

Parallel (bubbles in a scalar field)

Parental

Daughter

Negative and positive (matrix array)

Besides, moving between the levels makes us get the access to the same structure. Every level consists of the same.

Besides, negative and positive Universes may turn out to be nothing but an array of objects where all possible variants of objects are present, while negative and positive forms are just two outer extremities of the matrix array.

8.1171.

The scalar field of the Parental Universe where the bubble of our Universe grows in, as well as billions of bubbles of other Universes, greatly resemble computer memory where programmers once launched a virtual world and now some silly pupil plays this computer game.

8.1491. Void.

Void means pain, void needs to be filled, needs to get rid of pain. The Universe that was once created in the endless void, has an endless desire to fill itself.

Life can't stand any void, an aspiration for void means aspiration for death. On the other hand, void delights the eye. Void is a place for life. The place that some person and this person's thoughts can take.

8.1791. Philosophy of life.

Life is movement and light is what moves the fastest in this Universe. Light of mind, light of philosophy. Mind is philosophy. Philosophy is the basis of the mind.

8.2280. God's Desire.

Human desires are the only limitless entity in this Universe other than the Universe itself. This brings us to the conclusion that man's desires are the universe. The universe is God's desire.

8.3234.

The Universe is billions and billions of small boxes, nested one inside another.

8.3381. Awakening.

A human personality is very scalable. One day it's dormant, another day a point generates the whole Universe.

8.3385. On the question of equality.

Potentially, the point is the universe ...

8.3420. Antimatter.

The matter inert to time, existing out of time. The matter out of space and time.

Now they are trying to find an island of stability at the level of transuranic elements 113-120 in the periodic table, trying to go higher, increasing the core mass. But I think there still is the possibility to step aside, kind of slide aside, for example, new type of hydrogen - the one that could be in the other, parallel or negative Universe.

8.3560. A person covered with butter.

The Universe is egoistic as it does only what it likes and wants to, demanding people to serve its desires. But it likes loyal slaves who are ready to fulfil any of its desires. When the Universe loves you, it's a big happiness as this way good luck follows you. Health, youth, beauty, money, joy, happiness... if you behave yourself, they can even cover you with butter...

8.3580.

I deny such a thing as genius, any man is a potential genius if you let him get out of his point. The universe originated from a point, and any person is a potential genius.

8.3662.

The Universe asks for beauty.

The Universe begs on its knees...

The Universe draws its hands to you in hope...

And screams ...howls ...demands grinding its teeth with rage...

For you to make a squirm, go all out, but get more...

Beauty...

8.3686. The key to himself.

If you think you can't do something, you're wrong, you can do anything. You are the universe, folded into a point, inside you, like in a magic box, there is everything that you want. All you need is a key to yourself.

8.3739.

A point is as huge as the entire Universe. A point coming in between two people takes them infinitely away from each other.

8.3942. Watch yourself.

The fact that the universe is helping you can't stop you from messing things up yourself.

8.4654. Instructions from man and other intelligent worlds.

"Variothoughts" and the philosophy of Syntalism (quantum philosophy) are, in fact, instructions for developers of new universes and artificial intelligence systems.

8.4699. The paradoxical Universe.

Paradox is a variant of perpetual motion machine. Basically, paradox is the basis of the Universe's existence. Paradox is the endless source of motion energy.

If you break paradox, you will break the Universe.

8.4975.

All people are potentially equal, but not more. Somebody is still a point, while somebody else is already the whole Universe. The issue of equality is the issue of growth, somebody has already grown and somebody else is still growing.

8.5400. Healthy food of the mind.

In writing the texts in Variothoughts and other EA books, I follow the same strategy as Providence in creating our lives. I try to write what my readers like and what they don't like in the same way. I could write only "gingerbread", but the experience of 10 billion years of existence of this Universe says that this is wrong.

8.5615. So I decided to create the universe.

I'm angry because I'm sorely lacking in love. The lack of love in my body is so huge that it will fit the whole universe.

8.5790. A point is not an end, it is a beginning.

When you put a point, you should remember that our universe originated from a point.

8.5871.

It took only one moment to create the Universe. And it's being upgraded for 12 billion years already.

8.6048. Circle of life.

Fish quite happily and freely live in the aquarium, their memory is such that they instantly forget. Swimming in a circle, it seems to them that around all the time a new area, and the aquarium is a whole huge unexplored universe that can be endlessly studied...

8.6831. 13 dollars.

Let's assume that our Universe is a package information system - the game, that some young man bought in a shop on sale for 13 dollars. This young man came home, launched the system joyfully and now is enjoying the game where he is the king and god. Time passed by. The game is addictive, God has forgotten his job and personal life long time ago, the real world almost disappeared from his life, he hasn't been in reality for a very long time and why being there? In that world he is nobody, while here he is God.

8.6857.

The threshold of human stupidity can't be crossed, stupidity is endless. It's thought that our Universe is pure stupidity. We live in the endless ocean of stupidity.

8.7080.

Imagine if you were given a chance to live this life again, if you were willing to live it without me, without our children, without our conversations, our adventures, our travels... none of this would have happened...

For example, if I imagine this, I am seized with a chilling horror and bitterness, huge as the whole universe, bitterness.

8.7223.

Once I got a closer look and noticed that the Universe was very small and extremely similar to a point.

8.7229. The Universe is an illusion.

Illusions save man from terminal boredom. The reality is extremely small and similar to a point, whereas illusions make it more like the Universe.

8.7253. A virtual Universe.

Cryptocurrencies are a good example of work for work's sake, which is an infinitely increasing computing capacity without any benefit or justification whatsoever. From the standpoint of equipment producers and developers, it all looks very interesting, though. I dare suggest that one day all of these capacities will be converted into something more useful, for example, into a new virtual universe.

8.7648.

The very existence of the universe is a miracle. The universe is very orderly, the universe is order and order is a miracle.

8.7649.

A miracle is an order that man cannot create. In particular, man cannot create the Universe and he cannot create life, that is why the Universe and life are miracles.

8.7656.

The simpler an entity, the more energy its destruction releases.

...the universe was formed by the destruction of the most elementary particle.

8.7657.

The algorithm of life and that of the Universe are the same: it is an adaptable and ever-complicating algorithm. Initially, it created inanimate matter; it then became more and more complicated and, based on the latter, it created animate matter.

8.7754.

Can it be that relic radiation is time? All movement processes in the Universe occur in relation to it and time is a derivative of movement.

8.7979. One small chance.

A moment is a lot, you can create or destroy an entire Universe in just one moment.

8.8344.

The gigantic web of the Universe resembles a neural network is similar to the human brain... It's as if we all lived in God's head... Our entire world is God's fantasy and God's dream... We have a chance to live until he is interested in this fantasy.

8.8346. Zero that flees one.

If matter is considered as one and its absence as zero, it can be concluded that the more matter there is in the Universe, the bigger gravitational forces are and the more it strives to turn back into a point. Importantly, this point was the first ideal one that then started to be divisible by 3, 7 and so on in an attempt to compensate for the ever expanding mass of nothingness (zero).

8.8365.

Could time be compared to a pancake? Suppose, time is like a lens whose thickness and diameter keep growing. This growth is what characterizes the passage of time in relation to the expansion of space. Closer to the center of the Universe, time is the thicker

part of the lens, and it gets thinner and thinner along the edges.
8.8367.
I would like to note that the space of the Universe resembles a liquid (plasma) and matter in it spreads on the same principles as light in water. In fact, light and matter are one and the same, one of the States of matter.
8.8368.
Dark matter is the space of the Universe that is devoid of energy in some of its states.
8.8370.
Presumably, light in the Universe moves, most likely, not directly, but on some parabolic trajectories, just as it passes through the water column, reacting to changes in gravitational fields.
8.8371.
Dark matter forms some cellular web that divides space into cells. Let's call these cells memory sectors or units. Inside these sectors there are accumulations of galaxies. In fact, this is a cellular structure, intercellular membranes or an equivalent of the cells of living organisms. It can be assumed that the Universe is a living cell mass that keeps growing.
8.8373.
The structure of the Universe is perfectly visible through a microscope.
8.8374.
By the way the space of the Universe is not just the liquid and the hot liquid... plasma, i.e. state of borderline between gas and liquid.
8.8383. Speed of light is not constant.
Gravity changes from the center of the Universe to its edges, so the speed of light and acceleration of the Universe expansion as it moves away from its center is not a constant either.
8.8399. The flat Earth.
The Universe is flat in the sense that it consists of zeros and ones. The Earth is the same in this sense.
8.8403. Formating the hard drive.
Black matter is required to synchronize and structure matter (en-

ergy and information) in the Universe.

8.8407. Slow photons.

Most of the universe mass is the dark matter, but where does it come from? Suppose it's the stopped light. Stars burn matter in their reactors, turning it into radiation and light, when the light stops (still photons?), it turns into the dark matter. (Perhaps there is some cycle of energy).

8.8442. I'd like to see the dead Universe.

There's a curious thought: is there a web of the dark matter outside the Universe, which is gradually expanding within the space structured by the same dark matter?

8.8451.

Failure to understand the complex stems from misunderstanding of the simple. It seems to them that the simple is so simple that it's not worth paying attention to, they already know it anyways... Alas, it's a misconception. Understanding the simple is much harder than the complex. Inside, the simple is like a point, where the archived Universe lives.

8.8580. The expansion of the Universe is the process of its disintegration and self-destruction.

The universe is not growing, it is disintegrating. Eternal decay of matter. The big Bang was the act of starting the point to disintegrate and self-destruct. In an attempt self-destruct, she seeks to burn matter in vents stars, light this symbol of death and destruction matter.

8.8581. Just as the rest of the Universe.

A person's height may be said to be derived from his self-destruction. He grows by destroying himself.

8.8619. Information teleportation.

The technology that allows to encode a person in a digital form and reproduce him in digital space. Next, with the light beam the code is transmitted with the speed of light to another planet, for example, where people may be present in digital form. Upon arrival at the destination, the person can both remain in the information space and control robots, biorobots and other peripherals.

The code can be transmitted both with a light beam at the speed of light and with the quantum method of information transmission through the qubit system by quantum teleportation, the speed of information transmission by this method is totally unlimited.

Also such a concept as cracking the digital code of the Universe and uniting virtual digital worlds created by man with the body of our Universe belongs to this technology. Further, any object within these spaces can be moved only by changing information about its coordinates.

8.8638.

The universal point, array of all its states. This phenomenon is repeated at every level. Just above, in a living DNA cell, there is information about any of its possible forms and content. DNA accumulates billions of different variants within itself. And any occurring mutation is preserved forever. Once created, information is never lost. Even higher is the person as a whole, who also has enough information inside himself to become anyone, including a star or even a new universe.

PS. By the way, it's true that you can turn lead into gold, the only trouble is that it takes a lot of energy.

8.8639. The star is an idol, and the universe is God.

Becoming a star is not as cool as it seems, strive to become the universe.

8.8651.

Space is matter. Expansion of the Universe is the matter creating process. But, in fact, there is a process of creating the information that structures energy. During the big bang, energy was released and the process of creating the information that structures energy into space started.

8.8657.

Centeredness is a formula for success, love is the source of centeredness. When you love what you do, nothing can distract you from it.

8.8664.

Friends, we got advertisement instead of epilogue. Advertisement is the motor of buying and selling, we would die without advertising or would be unable to create anything. Please find below some descriptions and advertising of various projects created by Soloinc. It includes music, lectures, various kinds of service and investment projects. The Universe of Soloinc is the world where everyone can find something useful and interesting.

8.8884. Puff pastry theory.

Chaos movement is probabilistic in nature, but with the increase in chaos, Probability is gradually turning into Certainty. That is, increasing in size, chaos becomes order. But as the system is puff, order, increasing even more, turns again into chaos. And that goes on indefinitely.

9.1177. Malevich's Criticism.

Malevich is wrong... If the picture is a window into the world, then why is it square... The primary form is a circle. The universe emerged from a point, a point expanding into a sphere.

The beginning of art is a black circle. But since there are three truths in three-dimensional space and eight truths in eight-dimensional space, it would be fair to nullify art to a solid black sheet. Other versions of the truth are: a solid white sheet, a white circle on black, nested black-and-white circles in each other, nested rainbow circles, and even a white square on black.

The logic of having white here is that, on the one hand, white is the evolution of black on the way from Nothing to perfection, on the other, it is the abyss before the end.

9.1214. White is perfection, it is the end, and it is the beginning.

The beginning of art is nothing, but nothing in itself we can not show, because in the material world "nothing" is not. "Nothing" is the void in which the universe arose... But perfection and nothing are the same, so the white sheet is the beginning...

9.1705. Green leggings.

External beauty is needed to draw attention to the inner beauty. The outer is a subtle ephemeral form, the inner is a universe-wide entity. Beware of those who love only your form, remaining indifferent to the essence... It means they're not interested in you, and you have little to change.

9.1833. The interpretation of aphorisms. The depths of simple truths.

Aphorisms are the concentrated wisdom of life. The best minds of humankind have concentrated their best thoughts for centuries to put them into aphorisms so that other people would not fall into the same trap twice. Aphorisms are very useful, most of them should be learned by heart so that to be quick at taking decisions in difficult situations. But there's a problem about aphorisms: they only seem to be simple, thought the simplicity hides a great meaning that should be understood. Simple truths are the greatest truths of all. They are like dots that hide universes inside. Simple truths are the base of existence. Without understanding the simple, no one can understand the complicated. The book is a collection and interpretation of the main thoughts of humankind, the meaning of which should be known to any sensible person.

9.2668. Seeds of life.

The Universe emerged from a grain, rather than a simple point. This grain germinated, grew and yielded... One grain gave birth to thousands and millions of new, small universes.

9.2669. Diversity of universes.

Observations of nature show that seeds of life are different and even the universe itself may emerge from a grain and it may equally emerge from beans...

9.2974. A saving straw.

You can never neglect anything. The universe came from a point. Everything big has grown from small, and at the last second everything can be changed dramatically by a saving straw.

9.3156.

Can differently dimensional entities exist within the same universe?

9.3210. Disintegration of time into energy.

The volume of energy in the universe is constantly growing due to the disintegration of time into energy.

9.3214.

The Universe is a four-dimensional dot within eight-dimensional space.

9.3274. The algorithm of energy packing.

A space of any dimension is mutually transformable and able to fold in and out one another using these or those mathematical algorithms known as truth. Each space dimension has its own algorithm of energy packing: the lesser the space dimension, the greater energy concentration.

In a unidimensional and three-dimensional object, for example, in a unidimensional point and a three-dimensional universe there is an equal amount of energy, but it has been packed using different algorithms.

9.3289. Revisiting the energy conversion efficiency.

The energy conversion efficiency is 100% true only for the transformation of energy inside Newtonian space from one kind into another. In case of processes of energy packing and unpacking that take place between one- and three-dimensional spaces, there can be any whatsoever energy conversion efficiency. Inside a onedimensional point, there is as much energy as inside an entire three-dimensional universe.

9.3290. The evolution of the universe.

The universe evolves from a onedimensional to a multidimensional space by storing information. The exponential growth of information leads to a collapse of the mass under its own weight and to the system's production of critical mass and a supernova explosion. Specifically, there are two explosions: one comes outside and the other inside itself. The explosion inside itself produces a breach into Nothingness, a superconducting point, on the other end of which a new universe emerges, of greater dimensionality than the current one. Energy unpacks and flows from this three-dimensional world into a new seven-dimensional universe, thus creating it. The supernova gradually works off its energy in this world, becomes depleted and dies out, giving less and less energy to the new world. When energy is up, the new world, bereft of its source of energy, will collapse under its own weight and, folding back into a point, will produce critical mass, and everything will start over again.

9.3291. Who created the Universe?

There is no time in a one-dimensional space, that's why the system can create itself on its own.

9.4101. The Universe Soloinc.

They say that Soloinc is a star, but this is not true, Soloinc is the universe.

9.4383. The Universe.

Point is an object of infinitely small area and infinitely big volume.

9.4739.

There is much more darkness than light in this Universe, so the Russian word тьма (darkness) is the synonym of «many».. Darkness has other synonyms: stupidity, zero. There is much more stupidity than intelligence and zeros than ones in this system.

9.4808.

Very much energy is potentially hidden inside a fool. A fool is a point. An idiot is an elementary particle with a huge amount of superfluous energy hidden inside. For example, our Universe emerged from a point.

9.5022. Digital Gods. A concept of a bored God.

Digital Universe and its gods are the gods of a new world, virtual world, the world that is still in the process of being created by people. The world of games, books and movies. The world which god is a human being. The new world needs new philosophy and new gods. Now when a human being became god, he will have to reconsider many things and realize the inner state so that to look at reality from a new perspective.

9.5339.

Any foolishness is possible. Foolishness is a kind of superconductor. I won't be surprised if the very our Universe emerged because of a foolish accident.

9.5343. You don't have to cross the line, you better push it.

The very boundaries you have in your head are extremes. And Yes, going to extremes is a big risk. They tell you to push the boundaries all the time. In fact, they suggest you become a virus or cancer and grow indefinitely. While this may be correct, our universe is doing the same.

9.5405. Grey brain cells.

Our universe is certainly reasonable, however, people somehow remind me of parasites and viruses. No doubt, the universe is God. Now, what is his attitude to the parasites living inside him? This is a very interesting question, indeed. BUT !!! Maybe the people are not parasites at all. People are the God's brain – a sort of his neurons, his brain's neural network. People are the God's fantasies, his dreams and illusions.

9.5415. A piece of God.

God cannot be painted because God is everything and he is everywhere and always. You cannot portray something boundless. Wherever you look, God is everywhere. The Universe is God. God is a whole and he is truth. By painting a piece of God, you destroy truth, turning it into lie.

9.5523.

For a one-dimensional dot-man to evolve into a multidimensional system, he needs to engage in personal growth, overcoming his vices through pain and suffering. Creating the universe from a point is an explosion, a process of self-destruction and self-creation at the same time.

9.5546.

A one-dimensional person can become four-dimensional, like the Universe emerges from a point. He needs new software, four-dimensional tasks and great inner desire to do that.

9.5549.

Logically, all the elements of this Universe rotate on their axes constantly. This rotation starts from the very point from which the Universe emerged, is continued by the Universe itself and applies to atoms, electrons, stars, planets, galaxies and so on.

9.5642.

It's very interesting to explore inner world of some other person. A person often seems a point from afar, but if you take a closer look, he is the whole Universe.

9.5669. Searching for NOTHING.

Free space is necessary for a human to get out of his point and become the Universe.

9.5735.

Life is contagious, the normal state of the Universe is death, that is absence of life.

9.5956.

Point is the archived universe.

9.6260. Three main essences.

Einstein said that the Universe consists of energy and information. However, there is another essence in this system, and it's called FORCE.

9.6463. Desperate situation.

The situation when there is no way out is called a point. Indeed, there is no way out of point.

But you can push the walls. Remember – push the limits of mind and then the big bang will occur, the walls will disappear and a new universe will emerge.

9.6464. Broader perspective of circumstances.

There is really no way out of a desperate situation, that's why it's called desperate. But our universe is multi-dimensional, the same situation necessarily has thousands of its different variants. Don't fixate on what you have and choose the situation with a beautiful way out.

9.6891. Sex

In the three-dimensional universe objects have form, content and usefulness (the object's purpose and function). The object or process can be recognized by changing their form or content, but saving the essence. For example, sex between a man and a woman can be depicted as an electric generator and an electricity-generation process, a child can be added, pleasure can be depicted.

9.6892. Surprise me.

Art is an endless conversation with the universe, and strongly repeated and constantly pour from empty to empty is undesirable. Novelty and surprise are especially appreciated in this conversation. The infinite universe is endlessly bored.

9.7132.

Any chance in this universe is ruled by the Devil. Plans and order are under God's control.

9.7374.

The Universe will die when the quantity of information becomes equal to the quantity of energy. Originally, the quantity of energy in the Universe is a constant while the quantity of energy-linking information constantly grows. Originally there was energy only, later there appeared information and its quantity started to grow. What will happen if information structures all the available energy? Will the Universe shrink back into a dot?

9.7375.

Perhaps, black holes are some kind of space archivers, the mechanisms which function is to make the universe shrink back into a dot.

9.7635. The origin of a point.

A blank, white sheet of paper is the perfection of order. When perfection dies, it turns into chaos. A point results from the weakening of order and chaos' breaking through the canvass of reality. Further on, as perfection dies, a point turns into a new universe. Life is the movement towards death, from perfection towards chaos.

9.7658.

The dark matter in the Universe is something like bad loans and dead money accumulated in the economy.

9.7760.

Inside each elementary particle there is a universe. If such a particle is split, a big bang will occur and a new universe will emerge. A big bang is the process of transforming a uniform space from one dimension into another.

9.8073.

The Universe is objective. The point from which it originated is subjective.

9.8383. It is interesting to talk to someone who has something to say.

An artist is someone who speaks to the universe. But to speak to the universe, you have to have something to say to it. The universe will only talk to someone who is interesting to talk to, who will tell it something that no one has told it in the last 15 billion years.

9.8430.

A universe originates from a point and a man from an egg cell and a sperm cell. The question is, if the point was an egg cell, what was a sperm cell?

9.8745.

The Universe is, basically, a quantum computer system in which zero and one with a variable charge level encode information. The clock rate in such a system is the variable dependent on the speed with respect to the speed of time.

9.8746.

The expansion of the universe is the process of creating information. The more the clock rate, the faster the speed of the universe's expansion and the more information it contains. As the system grows, the dimensionality of space increases – one-, three-, seven-dimensional space and so on.

9.8747.

The endlessness of space has to do with its multidimensionality. The boundlessness of the universe has to do with the stepped structure of measurements.

9.8773.

Unidimensional space is a point, two-dimensional space is a plane and three-dimensional space is a sphere. The expanding universe is four-dimensional space.

9.8776. The curvature of time.

We know that is a two-dimensional space is stretched over a three-dimensional space in which the former will become form and the latter will become content, the two-dimensional space will be boundless because it will have no bounds. Boundlessness, however, is not endlessness because the area of such a space will remain a finite invariable.

That is, if a two-dimensional entity is given a three-dimensional form, it will be boundless but not endless. Correspondingly, if a three-dimensional space is given a four-dimensional form, it will also be boundless. Consequently, the boundlessness of our universe is due to the fact that it is a three-dimensional entity in the form of a four-dimensional one. It also means that a four-dimensional curvature is inherent in our three-dimensional universe. Considering that time is the fourth dimension in our universe, time in our universe can be said to be non-linear and curved.

9.8779.

The Universe is a three-dimensional dynamic sphere in which the fourth pseudo-dimension is a pseudo-invariable, or time.

9.8791.

Any uncontrolled process seeks chaos. A system can be said to strive from order for chaos and from one for zero. The normal state of a system is its destruction. The fact that a system is moving from one to zero suggests that the universe's existence is finite.

9.8792.

The speed of time gradually fades, time used to move faster in the past. In the future it will become even slower until it stops at all. The universe will die the moment time will stop.

9.8793.

The speed of time is a value reciprocal of the speed of the universe's disintegration.

9.8794.

The disintegration rate of the universe goes exponentially, the older the universe is the higher the rate gets, besides finally there will be a slumping growth of entropy.

9.8795.

The higher the disintegration rate of the universe is, the slower the speed of time. The higher the speed is, the slower time goes.

9.8799.

As the dimensionality of space increases, its area increases by the number of the primary system's elementary particles and the invariable of distance between them. In a seven-dimensional universe, a three-dimensional one looks like an elementary particle.

9.8800. The speed of light is not an invariable.

It can be assumed that the speed of light is related to the speed of the universe's disintegration and changes in proportion to it.

9.8805.

The Universe is God who created himself. So anyone, who was initially a nobody, can – by means of personal growth – creep out of his point and become a rather worthy person.

9.8853.

The biggest – and the smallest – digit in the universe is X.

9.8892. Man is the derivative of Y with respect to X.

The two major essences of the universe can be identified. The essence X is the array of all internal states whose value depends on external circumstances. The essence X is Content. The essence Y is the array of all external circumstances. The essence Y is Form. The essence Y is what is inside and the essence Y is what is outside.

9.8899. Energy is running out.

The amount of energy in the universe is said to be an invariable. At the same time, the universe is getting cold and the universe is disintegrating.

9.8901. Why does nothing flow into us?

The amount of energy in the universe is constantly decreasing. Black holes are the points of leakage of energy from the system through which our energy flows to other universes.

9.8917.

Time is the speed with which the universe grows cold.

... when the universe grows cold, energy will run out and time will be up.

9.9012. Mind games.

One day, we designed a trial self-learning artificial intelligence system that can simulate a game world inside itself. This self-learned game intelligence developed by playing with itself inside its virtual world. Fifteen billion years of internal game time had not elapsed when the SYSTEM managed to create an entire universe and a subsidiary intelligence inside itself. These creatures were called Humans, I guess.

9.9042. What does not exist yet.

The speed of light is related to the clock frequency of time in the universe. In theory, you can move faster than the speed of light, but then you will outrun time and will end up in the future, that is, in nothingness.

9.9329.

The speed with which the universe disintegrates is different in its different points. Consequently, the speed of time is different in the different parts of the universe.

9.9359. A strange construction.

If the speed with which the universe disintegrates increases, the speed of time increases too. However, if the speed with which the universe disintegrates exceeds the speed of light, the speed of time will start slowing down until time stops. Evidently, the maximum speed in this universe is the speed at which time stops.

9.9360. The universe stuck in a time loop.

The excessive speed of light results in the time starting to go backwards. That is, if the speed with which the universe disintegrates exceeds the speed of light, time will go backwards and the universe will fold back into a point.

9.9535.

The Earth may well be flat in a digital universe.

9.9846. An absolute point.

An absolute point is an entity that has neither volume nor area. The Universe originated from an absolute point and the absolute point is the smallest entity in this Universe. Any entity consists of absolute points. An absolute point can be folded in a point – an absolute point is an endless array of points. Any entity or even an entire universe can be folded in an absolute point. Any point can be placed within a point.

10.2047. There is no time.

Time doesn't move. Time stands still. Time doesn't exist. The universe is moving, galaxies and stars are moving ... man and his consciousness are moving.

10.2327.

The essence of infinity is scalability. A dot, a ball, and the universe are all the same.

10.3003.

Reality is energy and a source of energy. The unreality and emptiness is pure information, which sucks up the energy to give it form. A black hole is an entity that absorbs energy on the one hand and creates the universe on the other.

10.3044.

Life is motion, but there is no motion, because the universe is a point.

10.3495.

Since God is everything, then God is both a point and a universe. You can take any point and it will be God. If you don't want to take the dot, you can take whatever you like.

10.3501.

God is an absolutely scalable entity-a point and a universe at the same time. This is the nature of truth, love, and order. In fact, you can take any thing and love to scale it from a point in the universe.

10.3502.

The term "all" implies simultaneity. And point, and the universe, and lies, and truth, and chaos, and order.

10.3548. The theorem of the point.

Although various sources claim that the truth is incomprehensible, I do not agree with this. God is everything. God is the truth. All this means simultaneity and scalability. God is both a point and a universe at the same time. Therefore, by understanding the point, you can also understand the universe. At the time, Mr. Malevich drew a Black square, but to draw it, he needed 7 volumes of the theory of Suprematism. I want to draw a point and for this purpose I am already finishing the 10th volume of "Variothoughts". The main point of "Variothoughts" is to solve the point theorem by proving that a point and the universe are the same.

10.3549. God is everything.

Soloinc is the artist who drew the dot. A black dot on a black background. The meaning of the philosophy of Syntalism and the book "Variothoughts" is to prove the point theorem and prove that the point and the universe are one and the same.

10.3552.

The universe is a black dot on a black background, and there is no light. Light is the illusion of the mind.

10.3677.
God is perfect, but mistakes and uncertainty are part of perfection. The system is designed so that errors do not kill it, but generate novelty.

10.3682.
Truth is a solid state of energy. Matter in the universe is only 1%, but this does not mean that the rest of the universe is not.

10.3683.
Black matter is like a lie, it only tends to become ordinary matter, just as a lie tends to become truth.

10.3684. Solid world.
The density of matter in galaxies does not decrease when moving away from the center. That is, the density of matter, taking into account the dark matter is a fairly stable value.
"Are you thinking what I'm thinking?"

10.3685. Memory module.
The proportion of dark and ordinary matter in galaxies remains unchanged. Increases ordinary matter, dark becomes less. On the contrary, respectively, too.

10.3982.
The universe is patient, it does not wait for anything, so it is ready to wait indefinitely.

10.4264.
Thought is a self-organized form of energy. In this respect, thought is very similar to our universe.

10.4443.
According to Syntalism, the universe is a self-learning system of artificial intelligence, that is, the mind, whose goal is to create information, that is, to accumulate a knowledge base about all possible forms of matter. And yet, according to Syntalism, God is perfection, but it is the perfection of the future, not of the past. The perfect mind of the future creates itself of the past... And this God of the future welcomes all who make it more perfect, more diverse and more beautiful.

10.4541.
God is the universe that looks at itself through the eyes of man.

10.4544. Universe of mind.

Consciousness is the brain's fantasy of the real world. The brain, falling into pride, wants to create its own world and its own God within it.

10.4545.

The brain within itself creates a model of the real world, creates there an idea of itself and the Superego, the God of the new universe. Consciousness studies this model of the world. And on its basis makes assumptions about the real world, transmitting information to the brain.

10.4566. Over Ya.

The brain has created its own universe within itself, where there is the son-consciousness (I - Ego) and there is the Holy spirit "over Me". So is the real world, where there is the universe, man and the Holy spirit – the main law and the idea of being.

10.4576.

From the point of view of quantum physics, time does not exist, and there is no problem of eggs and chickens. And since God is everything, there is no problem of the first cause. Billions of years will pass and people, as a society, will become one whole God. We seek unity with God and with each other. Having achieved perfection and unity, God will create the universe anew, restarting the process in an endless circle. The point of re - creating the universe is that perfection is death, so when perfection is achieved, everything has to be erased.

10.4586.

In that moment, when God has made perfect, he became the universe.

10.4615.

God is the point. The universe is a lot of points. The universe is a way of self-organizing points. God is society.

10.4636. Computer simulator.

Inside, the brain creates a toy universe where the toy Self plays with toy copies of people from the real world. The subconscious, watching this game, makes some conclusions on the basis of which behaves in the real world.

10.4643.

How do parallel universes work?

- Just like different computer and virtual games.

10.4727.

God is the universe, which consists of perfect details. Perfect knowledge of one's own unique piece of reality transforms a person into a unique piece of God. And now, having achieved perfection in the small, a person can become a part of the big.

10.4736. Energy, truth and lies.

The meaning of Yin and Yang, truth and lies is that we live in a pseudo-three-dimensional world. There are three truths and one lie in this world. Time is a lie. Time does not exist. Time is the process of moving the three truths. But even three-dimensional truth is only information about the form of energy.

10.4764.

Since time does not exist, neither does the imperfect. Imperfection is simply perfection distributed over time.

10.4821.

What Ptolemy is that the heliocentric picture of the world, it's all idealism and pride. The world doesn't revolve around you or your idol. The world revolves around the center of the universe.

10.4868. The universe and the point.

God is perfection, and space and time are illusions that divide a perfect God into billions of pieces. The point of the game is to bring God back to the point.

10.4869.

Cybernetic Syntalism-quantum Nanophilography, which believes that time and space are an illusion that turns a perfect point into the universe. The meaning of our life is to collect the pieces of the puzzle, to collect fragments of perfection from them. The ultimate goal of the game is to fold the universe back to a point.

10.4870. The illusion of the universe expanding.

The amount of energy in a system, that is, the truth, is a constant. Energy is truth. However, the number of lies and illusions in the system is constantly growing, because time and space are lies.

10.4918.

Reality, as such, is not accessible to human consciousness. The brain has built within itself a copy of the universe with itself and is now playing with it. In such a situation, we default to believe that the truth is not available to us in principle, we can only approach it. A copy will never be equal to the original.

10.4932. Rational investment.

Chaos is the source of order. Chaos breeds an infinite number of the most different perfections. Order takes what it wants out of chaos and puts it into the order it thinks is beautiful and perfect. Chaos takes from order the fruits of its activities, growing at the expense of them. In fact, order is a kind of investment of chaos. I call this system a perpetual motion machine. The more order structures chaos, the more chaos it becomes.

10.4946.

Order is a machine that produces chaos out of chaos. And it is an amazing and magical machine that produces more than it receives. A machine that produces thousands of times more nothing out of nothing than it was.

- How can you magnify anything?

"Well, it's all an illusion.

10.5101. The energy cycle.

Slowing down, time turns into light, then into matter. The growth of the mother's mass under the influence of gravity causes a thermonuclear reaction, again dispersing energy and turning it into time and light.

10.5172.

Chaos is energy. Order is what makes chaos move not chaotically, but according to strictly defined algorithms.

10.5225.

Time and space are ways of organizing and limiting energy that turn Nothing into reality.

10.5531.

Time does not simply liberate energy from the bondage of form. Time is the energy that gets tired of being in the same form all the time.

10.5802.

The point is that the cause produces the effect, which, growing up, becomes the cause of its cause. This is the unity of Yin and Yang, good and evil, darkness and light... and even our universe was created by this function.

10.5971.

You cannot comprehend the universe without comprehending a grain of sand. After all, a grain of sand is part of the universe... Again same a grain of sand contained sand for attainments good those, that she a small and from it easier go down. A grain of sand does not blind the eyes, unlike all other idols who consider themselves superior to worms and maggots.

10.6179.

The point is also too much, only in the opposite direction. If a point is turned inside out, it becomes the universe.

10.6348. Solid-liquid essence.

The harmony of being is at the level of the Golden section, not the middle. A color world could be constructed with three colors, but reality uses four basic chemical elements. The fourth element makes the system invulnerable to errors. Pseudo hardness. A pseudo fluid. Usually the system is solid, but when the hardness is destroyed, the system becomes liquid and adapts to the destruction, restoring integrity.

10.6911. Black square.

As syntalist, I do not believe in mysticism, I believe in quantum physics, chemistry and Cybernetics. For me, someone else's soul is not dark, but a black box, a cybernetic artificial intelligence system. However, I see the universe and its God in the same way.

10.7136.

Our entire universe is essentially a mycelium. We are long dead and live underground, and the industrious ants, taking care of our mortal souls, bring leaves to our graves and feed the good mushrooms that gave us immortality.

10.7252.

At night I dreamed of a lot of universal wisdom, and by the morning I forgot everything... it's Annoying.

10.7340. The truth is the point.

It doesn't take many words to tell the truth. The truth is very simple and concise. Truth is as simple as a point where the whole universe is hidden.

10.7520.

The disease, having arisen, begins to build itself, it grows. Love, plants, and the universe behave in the same way...

10.7621.

Alcoholism, drugs, and other addictions generate pride, because they deepen dichotomous thinking. A person on the edge of a pre-cipice begins to feel like an unknown hero, confronting an entire universe of darkness and evil in solitude.

10.7634. The transformation of lead into gold.

Enlightenment can be divided into three stages. At the first stage, you realize that black is white and Vice versa. At the second stage, you realize that black and white are the same thing. In the third stage, you realize that all objects are the same, God is everything, and a grain of sand and the universe are the same.

10.7672. Restraint.

To contain in itself the enlightenment is a means to keep them-selves from having to fail to the point. A point is as huge as the en-tire universe. When you reach awareness and enlightenment, you begin to be drawn into any point like a black hole.

You look at a grain of sand and see the universe, you look at the door handle and see all the secrets of existence, you look at a leaf on a tree and see all the beauty of the world. This is all very tedi-ous and you need to find the strength to manage the focus of your attention, unnecessarily trying to ignore all the unnecessary de-tails.

10.7730.

Idolatry is the first stage of enlightenment, when you fall into a point and notice that it is the universe. At the second stage, you need to get out of the point and realize that there is a whole uni-verse of such points.

10.7731.

God is incomprehensible, because every grain of sand is a universe in which there is a whole universe of similar grains of sand.

10.7816.

The book of Syntalism is like a universe that grows from a point and decays into infinity.

10.7889.

The movement of galaxies and stars is a vortex process on the surface of a four-dimensional sphere, similar to hydro-flows in the atmospheres of planets. Given the relationship between time and space, we can say that time is nonlinear in its development.

10.8063.

The essence of three-dimensional space easily penetrates into two-dimensional space. For a two-dimensional entity, the three-dimensionality of being is incomprehensible in principle.

10.8243.

Time is a derivative of the velocity vector that depends on the current rate of expansion of the universe. In fact, the movement of galaxies and star systems is not an independent movement, but an artifact associated with the expansion of the three-dimensional plane of the universe sheet on the surface of a four-dimensional sphere.

10.8315.

Although God is everything and everywhere, we are saddened to admit that there is not universal wisdom in everything, and in many places there is only universal stupidity in its absence.

10.8325. Inside a black hole

It only seems as if our universe is growing from an internal explosion, rather, it is a black hole that draws in energy from the outside and therefore grows.

10.8326.

If our universe is a black hole, it also evaporates through Hawking radiation. In fact, we are talking about light leaving the limits of our universe.

10.8327. The universe within the universe?

If our universe is a black hole, then it turns out that there may be other black holes inside the black holes.

10.8389.

Syntalism is a philosophy of mind. According to Syntalism, the highest value of this world is the human mind. God is mind. According to Syntalism, truth is attainable, God is attainable, and the universe is attainable. The Trinity of Syntalism is Mind, Nothing, and the universe.

10.8411.

Light is not a two-phase wave system, but a three-phase one. Any wave, including time, must consist of three phases, not two.

10.8705. The desire to know God is greed.

God (being, life, and the world) cannot be comprehended, but it is absolutely necessary to know. And you can learn it very successfully. It is impossible to comprehend it, because it is huge. In a world that was created from a point, and where every grain of sand is the universe, knowing every grain of sand is a very problematic and very stupid idea. It will be enough to know the world around you.

10.8722.

Man is a little God. He is perfectly able to create something that did not exist before him... animals don't know how to do this, and man is a wonderful Creator. Not as cool as God, but he's still learning... he's still small. Will grow up so ...and he will start creating universes... he is already creating them now ...What are computer games, books, movies, but micro universes?

10.9070.

Universal pofigism is courage and humility, based on the knowledge that everything goes according to plan, and you should trust the divine Providence, trying to look for profit in everything that happens.

10.9097.

The universe knows how best and gives you what you need and when you need it. Your task is to take what is given and look for the benefit of it. The worst thing you can do is close your eyes in fear and try to escape from reality.

10.9290.

A mother and daughter who live together often fight, because the more unhappy a woman is, the faster a hero will be found who will want to save her from Baba Yaga. Women adore heroes who save them from all universal evil.

10.9316.

While all other philosophies spin to a point, trying to achieve simplicity, Syntalism has come out of the point and spirals outward. The point of this maneuver is to prove that the expansion will add us back to the point. That is, the disintegration of the universe is a process of simultaneous paradoxical folding it back to a point.

10.9512.

All suffering from pride and ignorance. The proud are the blind ones who were enlightened by Jesus in the Bible. The blind don't see the truth, don't see the system. Love is what gives insight. Love is mindfulness and knowledge. Love allows a person to see the real world, which is millions of times larger than it is seen by a person who does not have love. Therefore, the lover looks at a grain of sand and sees the universe, and the proud man sees in it a figure of three fingers... and says why all this is useless nonsense. The proud man looks at the lovers and calls them fools engaged in useless nonsense. The proud man does not know that he is blind and cannot see the priceless gift for which it is not a pity to give all that you have, and for which it is not a pity to die. That's why the deeds of the proud are small and like sand, and the deeds of lovers are huge, like huge worlds hidden in a grain of sand.

10.9674.

There was a motive ...and the motive was the cause. There was something else before this motive, and there was a third before it. And first of all there was a point from which the universe arose. But does this mean that this point had no motive to arise?

10.9758.

What is a symbol of beauty?

– Shrimp...

Why?

– I looked at it and saw the universal beauty in it... Yes, I know that there is beauty in everything, and I could have meditated on a grain of sand or a Lotus flower, but I didn't have any sand or Lotus at hand, and I needed energy. I had to extract energy from what I had.

10.9925.

Freedom is the rejection of desires and goals. Freedom is like an explosion. You used to be held back by your goals and fears, but now that they are gone, you become like the universe.

8.1472.4. The Universe of zeros.

The question is why make the system 0:1 more complex? Why break down Ane into 3 and 7 parts. Obviously, it is connected with the demand for accuracy of computation. Sometimes it's neither 0 nor 1, in other words the system considers any unclear number a zero. That's why this Universe has more zeros than anes. Many zeros are half-done anes. Ane is quite a perfect construction that has the right to lay claim to energy and time and that's why rigorous selection criteria are applied to it.

8.2127.2. As far as he's right, he's wrong.

Conclusion: do not take everything at its word, any point of view needs to be weighed. One point of view will be as heavy as the whole universe, and the other as light as nothing.

10.10157. Ellipsis is life.

Truth is dead. Truth is the point from which the universe arose. To get things moving, you need to turn it into an ellipsis.

10.10172.

It is very easy to control beauty, love, and truth... one should only manage one's attention by protecting certain entities and rejecting all others. However, at the level of the mind, you should know that these boundaries are conditional and you can move them wherever and how you want.

Beauty and truth is a complete Gestalt. The point is that Gestalt is a scalable entity that can be closed at the level of a point, circle, square, broken jug fragments, or blot... or even the entire universe.

10.10203.

The desire to narrow the boundaries, reaching the original truth hidden in the point ... is very similar to the apogee of template thinking. When your template is a point of view, you become extremely limited. On the other hand, given that all points are very similar, you can now scale your template to the size of the universe.

10.10213.

Maximalism is the desire to inflate out of a molehill and proclaim the point as God. The other extreme of maximalism is the desire to turn an elephant into a fly ... it Seems as if the universe is God, but God is not only the universe, but it is also God.

10.10214.

The final point of view is extremely small and limited... it is vanishingly small, it is not even visible in the microscope. This limitation is beautiful, full of energy and extremely similar to the truth... Moreover, it is the truth... but the truth is not a single point, the truth is an infinite number of points, which we call the universe.

10.10217. Book of life.

The book "Variothoughts" is imperfect. Perfection is dead, and the book "Variothoughts" is alive, it grows all the time and never stops in its judgments and conclusions. The book X "Variothoughts" has no end, because X is everything. However, life is not infinite, so at some point X will reach a critical mass and add up to a point, which will cause an explosion and the creation of a new universe.

10.10274.

Masturbation is selfishness and a disregard for someone else's joy. Masturbation is the embodiment of universal evil, which should be avoided in any form.

10.10301.

A clean white sheet is joyful. White is the joy of the killer who destroyed all forms. White is the joy of the Creator, anticipating the creation of a new universe.

10.10469. The expanding universe.

What do these gentlemen mean by unity? They created an idol and want us all to pray to the same idol. The idea is extremely unoriginal and has a lot of disadvantages. These people are not aware of the principle of simultaneity, the one is the totality of the many. God is therefore incomprehensible because the number of forms that make up him is infinite and grows faster than our ability to know.

10.10793. The trees and the forest.

Truth is huge as the universe, and there is a small as a moment. These two truths are simultaneously different and identical.

10.10808.

God is everything... both the point and the universe... both the tree and the forest... and the drop, and the ocean. The small and the large should be considered God at the same time. God is the universe and God is man. In any case, neither the universe nor God needs idolatry, but they need knowledge, care, and warmth.

10.10824.

The concept of "God is all" is revealed through the concept of the Trinity of unity of point, form, and picture as a whole. "Everything is God" is a point, a form, and the universe as a whole. And more than that, there is a goal that generates movement. Movement is life.

10.10996.

The Noah's Ark metaphor is a metaphor for the human soul, which contains original ideas. All the animals in the ark were pairs, which meant they were fertile and could breed a new population. The world of the human soul is a whole universe in which there is a place for all the pilgrims from the old world.

10.11306. Humility.

The most important task of Syntalism is to create the Holy Grail, a cure for pride and procrastination. The Holy Grail turns darkness into light, sand into gold, and slavery into freedom. The Holy Grail is a source of joy, energy, strength and health. What is truth? Truth is not a dot, but it is a ball, as big as our entire universe.

10.11423. The Carpenter Jesus.

The problem with our reality is that everything is unique... understand the desire to become an idol and the subject of praise. Everyone wants to become stars and creative natures. It should be understood, however, that love requires uniqueness and the perfect carpenter who loves his work is valued thousands of times more than hundreds of thousands of little-distinguished poets who want to become idols in their pride. Pride is the desire for uniqueness, to become an idol and earn universal love. Pride is when there is no love in a person, and therefore he greedily craves external love.

10.11515.

When you get to what you think is the whole, you will go around the circle and return to the beginning. Truth is a circle, the universe is a point. The universe begins with a point and ends with a point.

10.11579.

What is the matter? Some stable state of matter. Any matter, in fact, is energy and is capable of conversion to other types of energy.

10.11666.

The laws of the universe can be divided into three groups. Local laws of our world. Federal laws of the universe. The immutable and unchangeable Constitution of Being.

10.11861.

The whole is something that can transform itself into a community by reproducing its own kind. So the point becomes the universe, the tree becomes the forest, and the man becomes God. The Creator is the one who can create such creations that they can create themselves.

10.11896. The more at the bottom, the less at the top.

The dual two-sided integrity is like a pair of scissors. Large, capturing a lot of small, turns into a unit. At the level above, the big one again captures all units, turning them into one unit. At the top of the essence similar to the universe back into a point.

10.12137. Hunger.

The dot wants more than anything to become the universe. The point is infinitely lonely, there is not a drop of love in it. If you love her just a little, she'll explode with a lust for power and love.

10.12391.

Those who are courageous are not afraid of their surroundings. The point, crushed by the ocean of darkness, explodes and turns into the universe.

10.12451. Cognition of the point.

When you focus on a point, you will notice that the point is not as simple as it seems. The point is constantly changing. A point generates a stream of information. You fall into a point and realize that this is not a point, but the whole universe.

10.12472.

To focus your attention on a subject, you must use your will-power to discover new sides of it. In General, given that the point is the universe, this is quite possible.

10.12480.

Initially, any object is a point, but the more you know it, the more you open its edges, the more it begins to become similar to the universe. The larger the object, the easier it is to focus on it.

10.12635.

What you are looking to increase in size. Look at a point and it becomes the universe. Look at a man and he will become a God.

10.13047. Truth in depth.

They say the truth is deep... In the depths of what? In the depths of illusions? What happens if you pull it to the surface? Will it burst from the internal pressure? Isn't that how the universe came into being?

10.13080. Associative engines.

Interstellar travel is available in the same way as connections between objects in space/time, as well as connections between ideas in the human mind.

10.13083.

What is virtual reality? A conceivable connection between abstractions that allows you to instantly move within the universe of space/time, as well as outside it, visiting other universes.

10.13120.

The Creator must be huge in his plans, for the Creator is God. On the other hand, God is everything, so an infinitesimal point is identical to an infinitely large universe.

10.13130.

It's hard for me to read. I read very slowly, sinking into every line.

In every word I see the bottomlessness of the universe and the versatility of being.

10.13302.

In the human brain, next to each other, thoughts about objects that are significantly distant from each other in time and distance can instantly arise. A similar picture should be present in the real world. Movement in the universe is possible on the principles of associativity.

10.13368. A question of modesty.

It is pride to think of oneself as the center of the universe, and the same pride to think of the earth as the center. Even the sun is not the center of the universe. All of us, including our galaxy, revolve around the true center of the universe.

10.13369. What is our universe?

Time is a relative entity. Our universe is a nuclear explosion bubble that only exists for a few seconds, including all 15 billion years of our history.

10.13370. Four-dimensional space.

The big Bang is something that happened in time, not in space. They say there is no time, but I tell you there is no space, only time. Space is time in a four-dimensional coordinate system.

10.13373. Inside the light bulb.

The universe does not expand, but it becomes less dense, in fact, it is as if the fire burns out, devouring all the fuel, and gradually turns into light. Light flies out of the universe in an unknown direction.

10.13399. Nothing in common.

These people live together, claiming that they have nothing in common. What do they mean? The same drugs and addictions? On the contrary, these two keep each other from overdosing. If you find a related addict soul for the same sources of pleasure as you, you will very quickly exhaust all your joys, get used to it, and fall into a hellish drug withdrawal and universal depression.

10.13414.

The point is that there is not going anywhere, but time. You are a star in the ocean of space, and there are trillions of stars like you around you. Everything was still and silent. Only your mind wanders with its attention across the vast universe. All that you call "come, gone, love and fear.".. this is all an illusion held by your attention.

10.13634.

They say time doesn't exist. It's not like that. There is only time, we live in a stream of energy called time. Nothing exists except time. The three-dimensionality and movement of this flow creates space.

10.13735. Play, musician.

What you call the material world is rather a musical rhythm, a kind of melody that vibrates the single space of non-existence. Being is the music of non-being.

10.13845. The diameter of the truth.

No thought can unfold endlessly, but truth loops in a circle. However, the size of this circle can vary from a point to the diameter of the universe.

10.13878.

Mindfulness is the realization that a grain of sand today and a grain of sand yesterday are two different grains of sand. Every grain of sand is a universe, in the few minutes that you haven't seen it, millions of years could have passed inside this universe, changing everything dramatically.

10.13993.

Ignorance is when you look at a grain of sand and see in it the infinite beauty of the vast universe. Admiring this beauty, you proclaim it as your idol, start praying for it, start telling everyone that you saw God... Experienced people know that there are billions of these grains of sand. Everything is God, everything is beauty. It is ignorant to think that God lives only in the grain of sand where you saw it. God lives everywhere, even inside you. God is everything. God is a grain of sand. God is you. God is billions of other grains of sand. God is billions of other "you".

10.14434.

Love should be given not only to those who can appreciate it, but in principle to everyone and always. Love is the sun. The sun always shines. The sun doesn't run after anyone. The sun is busy with his work, directing his thoughts somewhere in the infinity of the universe. The idea that the Sun needs something from the Earth or other planets in the solar system is questionable.

10.14737. Illusion of the universe.

The existence of the universe is an illusion. There is only one absolute zero, and all other deviations from it in different directions only compensate for each other, leaving the existence of zero unchanged.

10.15143. Burn in me.

Creatures of the night come out to the fire to warm themselves in it. Creatures of darkness love fire. Fire, sensing fire, responds to it in return. But love requires restraint. Stupid moths are unrestrained, and for joy they throw themselves directly into the fire. The fire turns into a conflagration and rushes at those it loves, shouting: "Be mine, be like me! Burn in me! Burn joyfully! "-shouts the fire. But children of the night burn in the fire for some reason is not happy, moreover, it hurts them and all this universal love is more like a fiery hell than Paradise.

10.15191.

The sun is the engine of the solar system. The interests of the sun are directed somewhere in the abyss of the universe. So you, if you want to be the sun, find yourself a dream and strive for it with all your heart.

10.15389.

He is the truth who is not afraid to be a point. The one who, having become a point, can rejoice, is worthy of being the universe.

10.15445. Chains of suffering.

The great weakness and problem of the proud man is that he always seems to be disliked or loved a little. From such universal injustice, the proud man is always offended and falls into despondency, anxiety, the desire to throw everything and run, laziness and procrastination. In search of love, the proud man usually finds a drug, a pleasure, and is placed as his slave.

10.15609. Inner strength.

Truth is something that acquires the ability to grow independently from a point to the universe. Having gained inner strength, the point expands its boundaries and gains access to new energy within its boundaries. Having received this energy, it once again expands the boundaries, and so on to infinity. Truth is love.

10.15620.

The human mind was created by hunger. Pride is the human mind. When hunger wakes up, the mind gains the power of the gods. Soloinc, when writing "Variothoughts", deliberately brought himself to a state of absolute hunger. Feeling the universal hunger, the mind turned on and decided to take over the world.

10.15759.

Greatness is needed to be seen from afar, not to rise above others. Humility is what the universe is from afar, and a point is near.

10.15760. Point and universe.

The paradox of God is that the further you are from him, the bigger he is. The closer you are to God, the smaller he is. Wherever you go, God will always be visible to you, but will never loom over you.

10.15948.

Pride should not be reduced, but increased to a grain of barley. Small, like the one the universe came from. I am the vehicle of love. I am God.

10.15979.

A dream is both a universe and a point at the same time. To dream should go from small, but to large. You need to start with a thread and a grain, through light years, moving to the huge. The ocean begins with a stream, and thousands of other streams join the stream.

10.16210.

Once I noticed that I live on the planet Earth. Then I noticed that the Earth is just a dot compared to the infinite universe. Then I noticed that I myself was a dot compared to another dot.

10.16324. The third dream.

Having merged with God, we traveled with him through the uni-

verse, enjoying the contemplation of its beauty. An endless journey through an infinite universe. The infinite variety and novelty of beauty did not cause satiety, for nothing was ever repeated.
10.16347.
No... it is not the Creator who begets creation, but creation begets the Creator. It was not God who created the universe, but the universe created God, who then created it. Creation is a dream that calls the Creator to perfection, calling him to create himself.
10.16353.
Mind is the ability to see the small and present the large. Pulling a thread, you should already see a huge picture of the universe, woven from this thread. You should see a new forest in the tree. In a drop – the ocean. At the point, the universe.
10.16402.
The universe can be represented as a wave system extended in 4 dimensions. A kind of infinite number of growing soap bubbles embedded in one another.
10.16434. The universe.
The essence of meditation is admiration. You should meditate on the point, admiring its simplicity and perfection.
10.16463.
Love makes the small huge and the big small. Love is the power that creates the universe from a point.
10.16464.
The meaning of the expansion of the universe is knowledge. Since knowledge is infinite, the expansion of the universe is infinite.
10.16465.
The universe is huge, but love makes the huge small and the small big. He who knows love also knows the universe.
10.16484.
Truth is truth in any form. Truth in the form of the universe and truth in the form of the universe are essentially the same.
10.16585. Flash of light.
What is the universe? This is a four-dimensional soap bubble that expands. An infinitely fine line of pure white paper expanding in time.

10.16601.

We are three-dimensional beings living on the surface of a four-dimensional globe. In fact, our universe is a four-dimensional ball, that is, a three-dimensional plane. Wherever you go, you will always come back. Curiously, this ball expands, increasing its volume.

10.16604. Get to the point.

Once I took all my flaws and put them together in one whole point. A point is a synergistic operation that turns a point into the universe.

10.16616.

Less is more, but it is necessary to go to the point in the most difficult way, that is, through the creation of the universe and collapsing it under its mass into a point.

10.16617.

As the universe expands, it will inevitably reach a critical mass and form a point. This is the fate of all points and all universes.

10.16622. Point.

The word was in the beginning, and the word is truth. The meaning of "Variothoughts" is to create a universe of words, bring it to a critical mass and add it to a point, thus exhausting the words and comprehending the intuitive truth.

10.16673.

The book "Variothoughts" is necessary for training the sense of perspective in thinking. Truth is the unity of the many, the point within which the universe is.

10.16674.

The universe grows and becomes a point. Zero is the basic state of the universe, but when it reaches wholeness in the process of growth, it becomes a point.

10.16675.

The universe is not a ball, but a ray like a cone. Moving in space time, the universe forms a cone shape. In fact, it is a point source of light, which, as it moves away from its source, scatters and fades.

10.16676.

If we assume that the universe is a point source of light, then our reality is like a ray of light. However, the light shines in different directions, so there are parallel intersecting time processes with a displacement along the entire diameter of the original ball.

10.16719.

The universe is an entity distributed in time from 0 to infinity. It is important to remember that time is movement. Movement is faith. You just believe that time is running out. You believe that the universe is infinite. In fact, there is nothing. Nothing believed in itself and it created time and created the universe.

10.16826.

About the absolute, Syntalism defines God and the universe as an information system that accumulates information about the form of matter and its differences from the ideal. The ideal is a blank white sheet. Theoretically, we can say that a blank white sheet is an absolute. But a blank white sheet is an abstraction containing infinite terabytes of information.

10.16827.

The universe is growing and learning. That's right. But it grows not to the absolute, but from it, through the destruction of the absolute. The absolute itself is simple, it is a blank white sheet

10.16866.

The tree doesn't look very much like its seed, but it did give birth to its seed. So God is a point, and the universe is a tree.

10.16901.

The cybernetic philosophy of Syntalism considers the universe to be a cybernetic system, the purpose of which is to create and accumulate information about the forms of objects. God is the mind of this system, the algorithm by which information is processed.

10.16922.

The absolute is not one big star, but the universe. The power of the absolute is in the infinity of the starry sky, and not in the size of individual stars that think they are idols. There are asteroids, comets, planets, moons, stars small and large, black holes. Black holes are stars that have reached the limit of their mass and burned out, folded into a point. All these are parts of the absolute. The star does not approach the absolute, it is just a part of it. The absolute accumulates information about all its possible States.

10.16979.

Truth is not afraid to be not a point, not a universe. Truth is happy to be both a point and a universe.

10.17011.

Space is four-dimensional, time is the fourth dimension of space. The universe is a four-dimensional matrix.

10.17042.

To overcome pride, you need to realize that the universe and the point are one and the same. As long as a person is convinced that this is not the case, they should be called ignorant, liar, and proud. A proud man should burn in hell, that's fair.

10.17043.

To overcome pride, you must achieve perfection and realize the simultaneity of everything and everything. The point and the universe are one. No one is elevated, no one is belittled. We are white dots on a blank white sheet, and all our greatness is only the pride of illusion.

10.17123.

The greatest happiness is to be both ordinary, like everything, and special, like the universe and as a point.

10.17124.

To be, as it all means to be the universe. To be special is to be a point.

10.17216.

They are looking for simplicity to get to the point. Because the dot is the truth. Syntalism is like the universe, it grows from a point, searching for darkness and turning it into light. Those who look at the light shrink, those who look into the dark grow. Those who fly to the light are like moths who seek fire to die. Those who go into darkness are themselves fire.

10.17224. Variety of beauty.

Thought should not strive for perfection, but for diversity. The search for perfection produces a pure white or black leaf, devoid of beauty. The universe is growing in breadth, movement deep is limited.

10.17578.

Gutseriev's poems should be read as one reads the Holy books, searching for revelations and experience of knowing the truth in them. Each line makes sense. Each point is independent and self-sufficient. Every tree is a universe, but many trees are a forest that is also a universe. The universe is an entity hidden in a grain of sand of a flower and a Lotus flower. The poet of reality is the one who reveals reality to us step by step. Reality is unknowable, but here and there by distinguishing its details, we learn the true beauty and perfection of being.

10.17605. The power of love.

In its significance, Gutseriev's poetry is a revolution in modern art, of the same type as Warhol and Marcel Dushan, Salvador Dali and Mayakovsky, Pasternak and Jack Kerouac produced in their time. This is truly worthy of the Nobel prize in literature for discovering new forms, courage in art, and the triumph of the human mind.

Gutseriev's poetry once and for all defines the freedom of the human mind, freedom from patterns and circumstances, freedom from the dictates of the external world. For millennia, philosophers have asked the question: which is more primary – consciousness or being, internal or external. Gutseriev teaches us that consciousness is primary, love is primary, and the human mind is primary. By the power of love and reason, human consciousness redefines being. The mind is not a poor relative of circumstances, it can not only be equal in strength to nature, but it can also win in the unequal struggle of a small person against a huge universe.

10.17606.

They say that God created man, but what does this mean? This means that God created the mind, but who is this God? God is love, as all the world's religions teach us. God is beauty and truth, faith and order. Love creates the human mind. Love is the human mind. Reason and love are the same universe that created itself. A universe that is both a point and an infinity. The laws of love are such that it is impossible not to love love. Do good-religion teaches us that good is love. Where does love live? Love lives in the human mind, and love is the human mind. They say God created man. I tell you that man created God. But the truth is in the wholeness of God and man. Mind is the divine.

10.17612.

Objects seem large, objects seem small, it's all a lie, it's all an optical illusion. All objects are similar, each object is a simultaneous point and universe, being and not being, potential and reality.

10.17619.

There are 69,000 texts in "Variothoughts". Soloinc was born on March 1 under the sign of Pisces in the year of the monkey. 69 this is the symbol of the fish, the fish is the symbol of Jesus, the symbol of Christianity. Fish is a symbol of the unity of illusion and reality, truth and lies, good and evil. Syntalism is the philosophy of love, the philosophy of unity and integrity, the philosophy of harmony of the human mind and the real world. The most important mystical number of Syntalism is 169, the essence of it is that the point and the universe are one. A forest is a multitude of trees, a unit within itself is infinite. Another mystical number Syntalism is 8, in fact, 8 is the unity of 69.

10.17707. Why would you?

If the universe is a complex mechanism, then everything inside it is its parts that perform certain tasks. Answer me, detail, why are you here?

10.17769.

The growth of the entropy of chaos is not the destruction of the system, but its striving for perfection. The growth of chaos is the growth of layers of order in the system. The order strives for power and control. The more order, the more chaos. Restraint of order reduces the level of chaos in the system.

10.17831.

Man creates himself through habit. The self-creating force of habit is the truth by which the universe creates itself.

10.18251.

What is eternity? A moment of love. The universe is huge for those who have no love, but for those who have love, the universe is a point.

10.18503.

What is the universe? 1% of light in the realm of infinite darkness?

10.18504. A flash of light.

I noticed that the density of the universe is falling, it seems to be fading.

10.18508.

Nothing came up light. Light gave birth to time; time consists of space. The slow light is converted into matter.

10.18667. Graphene the universe.

We live in a universe of lies that came out of nothing. Nothingness gave rise to existence, the lies created the truth.

10.18669. First there was chaos.

The universe, business, and all other good things are created in two stages. The first step is to find or create chaos. In the second stage, this chaos is structured and ordered into a system.

10.18670.

Order requires chaos. Order is a form of systematization of chaos. On the other hand, chaos is just a simplified linear order. Subsequently, more and more complex system rules are imposed on it, which turns it into a super-chaos.

10.18695.

The universe is infinite, but limited by time. Time is looped.

10.18703.

If a lot of" nothing» put it in a big pile and arrange it, you will get something very interesting and similar to graphene.

10.18868.

A point is an entity outside of time. The universe is a point distributed in time.

10.19143.

The universe does not grow outside of itself, but within itself. The universe is a point that is boundless within itself and grows infinitely by division.

10.19190. The universe is truth.

The essence of the universe (truth) at the same time, it grows through the growth of elementary entities (energy) outside and through the division of the main entity into various forms within itself. Competition of forms for energy (content) occurs through uniqueness. Different unique forms can control the same energy without creating contradictions with each other.

10.19191.

The book Variothoughts, in fact, is an analog of the universe, an analog of truth. Variothoughts grows endlessly outward by increasing the number of texts in the book (more than 80 thousand). Texts are analogous to energy and content. At the same time, the book Variothoughts grows inside itself, by creating various new forms (Variothoughts books, more than 700 pieces). Forms compete with each other for content, but this competition is not directly, but through the uniqueness and usefulness of forms. The same content can belong to dozens of different forms.

10.19196.

If we compare the universe to a living being that eats and grows, then the size of truth is limited by the limits of its growth. On the other hand, it is said that the whole can divide infinitely within itself.

10.19197.

The density of the universe is falling because the amount of energy in the system is constant, and the universe itself is expanding and dividing.

10.19308.

The universe is a point that grows within itself, so the source of its energy is inside it, and it does not compete with anyone and does not waste energy on war.

10.19315. Step into the abyss.

What you call the destruction of the system and the growth of the entropy of chaos is an amazing thing when the system itself grows. In fact, we are seeing a fall into the abyss, a system of units begins to divide within itself and create new forms. The avalanche-like explosive growth of the number of these forms seems to you to be chaos, but, in fact, it is many different layers of order. The greater the depth of separation of forms in the system, the more such a system generates energy.

10.19351. Being is a reflection of non-being.

There is nothingness. This Is God. There is a devil. This is the one who is the conductor of ideas and the Creator of being. Being is a reflection of non-being, a version God in space Being. Being is the material world and our universe. There are an infinite number of such worlds, but there is only one non-existence.

10.19530. Absolutely black.

Black is the evolution of white. The perfection of the whole, breaking up into innumerable forms, becomes absolutely black. White is just one form, and black is our whole world of forms and illusions. Black therefore has zero energy charge, which gave all the energy to being. The absolute black.

10.19582. The principle of the big Bang.

Everything in our system works on the same principle, which can be called the principle of a controlled chain reaction. The system, striving for perfection, reaches a critical mass, and begins an avalanche of explosive decay and growth, which must be restrained and diverted excess energy. So the universe works, stars, society, government, business, people, knowledge, etc.

10.19669.

A black hole is something that, on the one hand, absorbs the energy of our universe, and on the other, divides it and structures it into new forms, creating a new universe.

10.19747.

There was no big Bang. The universe was a point, and it remains a point. We all live in a point. On the one hand, the universe is a division of a point within itself. On the other hand, the universe resembles an hourglass, where division and growth inwards generate growth outwards.

10.19749.

The universe expands not in space, but in time, where time is a stream of unstructured energy. Space is structured energy.

10.20367.

There is no struggle for a place in the sun. Point already exists. In the future, accumulating information and reaching a critical mass, the point begins to decay inward, so turning into the universe.

10.20415.

There is no first or last, everything exists simultaneously. The universe is a point.

10.20695.

Nonbeing is energy, and being is what gave it form. Non-existence is primary. First there was energy. When the energy reached a critical mass to avoid an explosion outside, it broke up into forms inside itself. The greater the amount of energy per volume, the more forms there are inside it. A point is a reference energy-intensive object that creates the universe within itself.

10.20696.

The more complex and ordered an object is within itself, the more solid and large it is outside. Hardness allows you to maintain the integrity of an object as its size increases. The more complex an object is, the greater its mass and gravity. At the same time, the complexity and number of forms within the object generates unity and struggle of opposites, warming up the object. The higher the pressure inside the system, the hotter it is. As the quantity increases, the quality of the object changes. As the object grows, it becomes an asteroid, satellite, planet, and star.

10.20697.

Inside the more orderly and diverse object, the hotter it is, stronger, has great power and produces more energy. The more complex an object is, the more contradictions it has. Contradictions are a source of energy. Order and humility allow, while maintaining contradictions and even increasing them, access to internal sources of energy, while avoiding the collapse and destruction of the system.

10.20702.

Chaos is primary, because as the mass and volume of chaos increases, its internal pressure increases,and the system begins to structure itself into order. The disintegration of chaos within itself generates a variety of forms. Forms, this is the order.

10.20705.

The greater and more irreconcilable the contradictions within the paradox, the more it all looks like truth, beauty, life, love, wholeness, stars, and the universe.

10.20711.

Truth is something that has boundaries within itself, not outside. Truth, having exhausted the concept of form, destroyed all external boundaries and rushed to infinity. At the same time, truth has created thousands of new forms and entities within itself.

10.20712. The truth is love.

Truth is limited on the outside, but unlimited on the inside. At the same time, truth is boundless on the outside and limited on the inside. Within the truth there is an infinite variety of forms that fight for its content. Truth is a matrix array containing an infinite variety of forms within itself. At the same time, outside of itself, truth is something similar to the universe. When we look at truth from the outside, we see one infinitely vast entity within itself, with an infinite variety of forms and contradictions.

10.21233. Eraser.

Let there be light! The creation of light was an act of destruction of darkness, the further process of disintegration of light is an act of creation of darkness. Perfection is darkness. That is, in fact, the creation of the world began with taking an eraser and erasing a part of nothingness to make room for a new creation.

10.21525.

I think that information and energy are the same thing, only from different sides. The more information you have, the more energy.

The more energy, the more information. There is no information without energy and energy without information. Only the forms in which the energy resides change. Money is the simplest form of energy. Things and people are more complex forms. The more perfect the form, the more energy is concentrated in it.

10.21869. The framework of being.

Black matter is something like the concept of a crystal lattice known from chemistry. Black matter gives shape to reality.

10.21880.

The absolute is chaos, and chaos is superorder. Thus, the absolute is darkness, and light is pure energy, not yet structured by forms.

10.21892.

The universe is an infinite ocean of energy, but it is virtually divided into borders, conditionally called their place, time and forms of things.

10.21919.

The universe arose from a point and everything is that point. You are also part of this point, which is everything.

10.21920.

You don't need energy to create the universe, just a simple quantum fluctuation.

10.21921.

The speed of expansion of space is much faster than the speed of light.

10.22009.

If space can stretch at an infinite rate, then an infinite rate of growth in the length of time is also possible.

10.22052. Inside the point.

Forms are equal to each other, just as illusions are equal to each other. The content is just one point divided within itself into an infinite number of forms. The power of forms over each other and the point is ephemeral.

10.22107. Puddle in the rain.

Reality can be imagined as a puddle into which the rain falls (a stream of time energy), filling with energy, the puddle spreads more and deeper, thus turning into an ocean.

10.22108. A spreading puddle.

The universe is an ocean of energy, into which there is a constant influx of time energy, from which the universe grows, spreading like a puddle.

10.22109. The cycle of time in nature.

We can assume that the universe is an energetically closed ocean of energy, where the future is a stream of energy previously evaporated from the ocean itself. Thus, the process of increasing the universe is a situation where evaporation turns into the future and then returns to the present.

10.22110.

To a greater extent, the expansion of the universe takes place in the past, but the deepening of the system inwards generates a proportional growth in breadth. The deeper the system, the more complete and larger it is on the outside.

10.22137.

Time and space are one. The process of creating the universe was associated with the stretching of space, therefore, there was a parallel stretching of time. We can say that stretching in time is the process of disintegration within the system, and stretching in space is its expansion outward.

10.22138.

The situation of stretching space and time is a situation of simultaneous growth of the system in three dimensions.

10.22139.

If we assume that the stretching of space and time are related to each other, then time is also a constant that cannot be changed. On the other hand, it is possible that time is a multiple process when many parallel threads are possible. By the way, there can be many parallel streams of space.

10.22140. Hourglass.

The stretching of space has created the stretching of time, but it seems that we are observing a closed process with negative feedback. At the moment, we see that the universe continues to expand, and time, on the contrary, collapses as the future turns into

the present, while the size of the universe grows. In fact, I see the mechanism of an hourglass. At the first stage of the explosion, there was a stretch in time, when it reached the limit and began to fold, this gave rise to a stretch of space. In the second stage, when time runs out and the universe reaches its limit, the reverse process begins, the folding of space and the expansion of time.
10.22141.

In fact, time and space is the fact of the rebirth of one-dimensional space into its three-dimensional projection. The explosion, as a fact of the creation of the universe, was associated with the stretching of one-dimensional space time, which gave rise to its three-dimensional projection. Similarly, we can assume that the matter is not limited to three-dimensional echoes, for sure there are seven-dimensional spaces, etc.
10.22143.

The deeper the system is in the past, the longer it is in the present.
10.22144. The concept of the hourglass.

The system, when viewed in time projection, reminds me of an expanding cone. In other words, we observe that as the system is stretched in time, it is stretched proportionally in space. The system expands from the past to the future and expands simultaneously in space. We can assume that there is no future, the flow of energy comes from the past, and it seems to fall down.
10.22154.

Time is the flow of pure energy, and the material world is the information that structures and shapes this energy.
10.22161.

The space of the universe is stretched, and the space of time is folded back to a point.
10.22162. The ongoing big Bang.

There is no expansion of the universe, there is a process of stretching the space of the universe. Proof of this is the decrease in the density of matter in the universe. In parallel, there is a process of adding time to a point, expressed in the process of accumulating past time and concentrating it into a single entity.
10.22182.

The future is the decay of the energy of time, which generates a proportional expansion of the universe.

10.22205. How universes are created.

Business is essentially a temporary quantum fluctuation. First, we invent money (or borrow it), then we use it to do business and create material values, and then we pay off debts or destroy the invented money.

10.22207.

The universe was born out of longing. God was bored and created the universe.

10.22213. Negative world.

We live in a negative world, because darkness is perfection, that is, information that gives form to matter. Light is pure energy, devoid of forms. Light is a stream of energy, time, the future, non-existence.

10.22216. The division of an atom in time.

By taking a lie and stretching it out over time, we will create the truth. However, the smaller the original grain, the more energy will be released from it during division. In fact, our task is to create a chain reaction of decay in time.

10.22363.

Stretching the point space to the size of the universe caused the release of energy, which became the basis of the energy and matter of our world.

10.22365. The pulsating universe

Our universe is pulsating, it is decomposed from a point to the universe, then it is folded back, and each time it is decomposed into the universe, additional energy is released.

10.22395. Infinite love.

Truth is reality, and reality is energy distributed over time. An illusion is an energy locked in an instant. The energy of the moment is huge, but to turn it into truth and the real world, you need to destroy the moment and turn it into infinity.

10.22454. Unity of the point and the universe.

What is the difference between a seed and a tree? In fact, it is the same extended in time. The forest and the trees are uniform in time. In the same way, all people and the entire universe are United in time.

ABOUT THE AUTHOR

8.2479.

SoloINC (anc.greek "combining the uncombinable", keeper of the grain")

Soloinc Logic, philosopher from the city of Sofia. Soloinc (Diamond Solo / Solodilov Dmitry), Bulgarian psychologist and Stoic philosopher. Supporter of the merger of logical and sensory methods of cognition. He considers the connection of traditional philosophies with modern science. He is the founder of the cyber-philosophy of Syntalism (Quantum Nanophilosphy), which considers the problems of philosophy, sociology, psychology and economics in terms of systemic cybernetics and logic.

Soloinc is not the first, but the last philosopher. Evangelist and cyberpunk guru. The author of more than 73 thousand original ideas and thoughts. Main books: "Variothoughts", "Diamond Stoic", "Theory of Existence", "Money Bible", "Quantum Philosophy", "Mathematics and Progression", "Velerechie", "The Device of the Mind", "Royal Fool", "Liberastia" , "Surrotic", "Surfutur" and others, in total more than 888 books.

3.1753.

In fact, Variothoughts is very tedious. I have seeked the truth all my life, then I found it and concealed it in a different place. Variothoughts is an intellectual quest and a mosaic of truth, broken into thousands of pieces. I found the truth in plain sight and concealed it back as well as before... What's the point? It's a game or a way to kill boredom. We live eternally and boredom turns our life into hell. I want to save you from sufferings for some reason...

10.21128. Soloinc Music

Soloinc Music is a stunningly beautiful integrity of music and text, admiring metaphors and secret meanings. Soloinc Music is a pleasure for living minds who have dedicated their lives to the search for beauty and truth. Soloinc Music awakens the minds and ignites the heart. Everyone will find joy and strength to live in it.

10.2341. A realistic mysticism.

The genre of poetry and music of Soloinc is a mystical realism. Most Soloinc songs are mystical ballads or religious hymns, prophecies, and insights. Soloinc lyrics are always metaphors and mystical signs. They cannot be taken literally. These are grains of sand in which entire worlds are hidden. All words are the opposite. To understand the meaning of the Variothoughts texts you need to read from bottom to top, from right to left.

SYNTALISM - GENERATIVE QUANTUM NANOPHILOSPHY

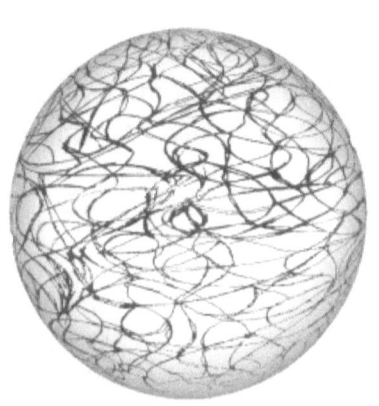 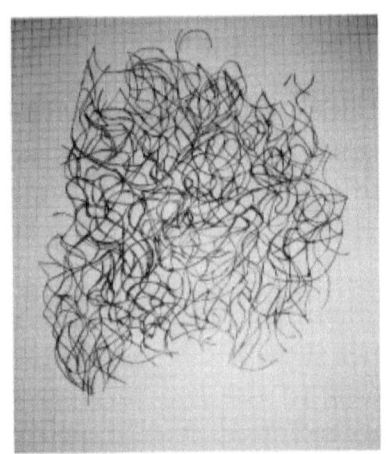

10.19296.

The philosophy of Syntalism was inspired by the poetry of life, expressed in the poems of such poets as Shakespeare, Robert Burns, Williams Blake, Pasternak, Lermontov, Mayakovsky, Velimir Khlebnikov, Paul Eluard, Andrey Bely, Alexander Blok, Voznesensky, Asadov, Gutseriev, Anna Akhmatova, Tsvetaeva, and others. Where if the philosopher had not come, the poet would have been there. Poets are like rays of light showing the way to thinkers.

5.782. Syntalism is the philosophy of the 5G generation.

Small thoughts are the philosophical system built in the millimeter wave range. Syntalism is 5G philosophy in the millimeter wave range built according to generative genetic algorithms.

5.767.

In Variothoughts, conceptualization follows the generative genetic algorithm.

5.768.

Variothoughts is the self-teaching guide on generative philosophy.

10.22348.

Syntalism is a philosophy that connects the unconnected with the goal of achieving integrity. Integrity is truth. To know the truth, the mind must cultivate tolerance and humility.

5.783.

Variothoughts is structured as a phased antenna array that ensures a dynamic horizontal and vertical growth of thought according to the generative algorithm and makes it possible to create different-sized logic data arrays. This solution minimizes energy consumed to maintain the integral information field. Variothoughts is a system of small cells in the millimeter wave (super-small thought) range in which the size of cells and their interaction structure are dynamic in nature.

10.3109. Unified system of knowledge.

The philosophy of Syntalism is by far the most perfect and clear philosophy, revealing the nature of being. Syntalism is like an ocean containing all other philosophies and religions. Syntalism understands and explains any point of view, agrees with everyone and loves everyone, considers everyone beautiful. Thousands of points of view, uniting into streams and rivers, turn into an ocean of Synthism.

VARIOTHOUGHTS COLLECTIBLE BOOKS

4.3423. Sand vs truth?

A book's collectable from the Variothoughts series costs only 1 cubic meter of real estate property. It is a very delicious price for something priceless.

3153.

God loves collectors as they give work to many creators...

6.6033.

The electronic version of Variothoughts is huge but printed versions are more complete and this book's collectables and handmade versions are unique in their completeness. Each of the author's gift manuscripts of Variothoughts is handmade and customized, that's why it includes even the latest texts that exist only in rough copies and have not yet been published anywhere.

10.22513.

Friends, I have not sold any Variothoughts collectibles yet. Pride rules people, that is cowardice and greed. There are very few courageous and intelligent people. He who is brave and buys the first book is very lucky. The first collector's copy of Variothoughts is a great value. Each collection book is registered and numbered. However, there will never be many of them, if I sell such books at least a few pieces a year, it will be good.